Jörg Fischer

Optimal Sequence-Based Control of Networked Linear Systems

Karlsruhe Series on Intelligent Sensor-Actuator-Systems

Volume 15

ISAS | Karlsruhe Institute of Technology
Intelligent Sensor-Actuator-Systems Laboratory

Edited by Prof. Dr.-Ing. Uwe D. Hanebeck

Optimal Sequence-Based Control of Networked Linear Systems

by
Jörg Fischer

KIT Scientific Publishing

Dissertation, Karlsruher Institut für Technologie (KIT)
Fakultät für Informatik, 2014

Impressum

KIT Scientific Publishing

Karlsruher Institut für Technologie (KIT)
KIT Scientific Publishing
Straße am Forum 2
D-76131 Karlsruhe

KIT Scientific Publishing is a registered trademark of Karlsruhe Institute of Technology. Reprint using the book cover is not allowed.

www.ksp.kit.edu

Print on Demand 2015

ISSN ISSN 1867-3813
ISBN 978-3-7315-0305-7

Optimal Sequence-Based Control of Networked Linear Systems

zur Erlangung des akademischen Grades eines

Doktors der Ingenieurwissenschaften

von der Fakultät für Informatik
des Karlsruher Instituts für Technologie (KIT)

genehmigte

Dissertation

von

Jörg Fischer

aus Göttingen

Tag der mündlichen Prüfung: 15.07.2014

Erster Gutachter: Prof. Dr.-Ing. Uwe D. Hanebeck

Zweiter Gutachter: Associate Prof. Daniel Quevedo, Ph.D.

Acknowledgments

The results presented in this thesis have been obtained during my time as a research assistant at the Intelligent Sensor-Actuator-Systems lab (ISAS) at the Karlsruhe Institute of Technology (KIT), where I have worked in the Priority Program 1305 "Control Theory of Digitally Networked Dynamical Systems", funded by the German Research Foundation (DFG). During this time, I was supported by many people. Among these, I would like to express my gratitude especially to those mentioned below.

I would like to thank Prof. Uwe D. Hanebeck for the opportunity to work within the Priority Program 1305 at ISAS and his supervision of my thesis. Certainly, this work would not have been possible without his continuous support and encouragement. Furthermore, I am very grateful to Prof. Daniel Quevedo for co-advising me and for coming all the way from Australia to Karlsruhe for the doctoral examination.

I also would like to extend my appreciation to all the current and former research assistants at ISAS: Markus Baum, Frederik Beutler, Christof Chlebek, Florian Faion, Igor Gilitschenski, Vesa Klumpp, Gerhard Kurz, Benjamin Noack, Ferdinand Packi, Antonia Perez, Marc Reinhardt, Patrick Ruoff, and Jannik Steinbring for the technical discussions and the great time we had at ISAS and Karlsruhe. In particular, I would like to thank Achim Hekler, not only for the impromptu assistance, joint work, and numerous contemplative dialogues about control theory and the world, but also for the refreshing implementations of the airborne rescue scenarios. In addition, special thanks go to Maxim Dolgov for the excellent collaboration, the eye-opening discussions, and his patience with my newest ideas on how to modify our entire simulation framework.

Last but not least, I would like to thank my parents and my girlfriend Susi for their encouragement and their belief that my doctoral studies would eventually come to a happy end.

Freiburg, October 2014 Jörg Fischer

Contents

Zusammenfassung

Digitale Datennetzwerke werden in der Regelungstechnik bereits seit mehr als 20 Jahren eingesetzt, um die verschiedenen Komponenten eines Regelkreises (d.h. Sensoren, Aktoren und Regler) miteinander zu verbinden. Klassischerweise werden dazu spezielle proprietäre Netzwerke, sogenannte Feldbusse, verwendet, die bei ausreichender Dimensionierung eine deterministische Datenübertragung mit garantierter Latenzzeit gewährleisten. Häufig ist es allerdings wünschenswert oder sogar notwendig Drahtlosnetzwerke und/oder nicht-proprietäre Netzwerke wie das Internet anstatt von Feldbussen einzusetzen, da diese nicht nur kostengünstiger und flexibler sind, sondern auch die Realisierung gänzlich neuer Anwendungen ermöglichen.

So eröffnen Drahtlosnetzwerke nicht nur im Rahmen der Fahrzeug-zu-Fahrzeug Kommunikation neue Möglichkeiten der Verkehrsfluss- und Fahrzeugkolonnenregelung sondern können auch in der Fertigungsautomation kostenintensive Signalübertragungselemente überflüssig machen. Zudem lassen sich mittels des Internets kostengünstig große Distanzen überbrücken wie diese beispielsweise in der Telerobotik auftreten.

Der Einsatz solcher Netzwerke stellt allerdings eine erhebliche Herausforderung dar, da diese Netzwerke kein deterministisches Übertragungsverhalten aufweisen. Insbesondere können, anders als bei deterministischen Feldbussen, starke Schwankungen der Latenzzeit sowie erhebliche Übertragungsverluste auftreten. Bleiben diese Störeffekte unberücksichtigt, ist mit einer massiven Beeinträchtigung der Regelgüte zu rechnen, die bis hin zur Instabilität des gesamten Systems führen kann.

Innerhalb der letzten 15 Jahre konstituierte sich daher mit der Disziplin der vernetzten Regelungssysteme (engl.: *Networked Control Systems*) ein Forschungsbereich, der ebendiese Problemstellungen im Schnittpunkt von Regelungs- und Kommunikationstechnik untersucht. Dabei wurden zahlreiche Methoden entwickelt, um störende Netzwerkeffekte bereits während des Reglerentwurfs berücksichtigen zu können. Eine dieser Methoden ist die sogenannte *sequenzbasierte Regelung*, deren Analyse, Erweiterung und Anwendung Gegenstand dieser Arbeit ist.

Die sequenzbasierte Methode nutzt die Eigenschaft moderner Netzwerke wie z.B. Ethernet-TCP/IP, dass Daten in Form von Datenpaketen versendet werden, welche gewöhnlich mehr Informationen transportieren können als benötigt. Ein sequenzbasierter Regler macht sich die freie Datenkapazität zu Nutze, indem er zusätzliche Informationen in Form sogenannter Stellwertsequenzen überträgt. Mit Hilfe der Stellwertsequenzen können die Auswirkungen möglicher Verzögerungen oder Verluste nachfolgender Datenpakete auf der Empfängerseite effektiv kompensiert werden.

Den Kern der Dissertation bildet das im Rahmen dieser Arbeit entwickelte sequenzbasierte S-LQG (Sequence-Based Linear Quadratic Gaussian) Verfahren. Das Verfahren vereint die sequenzbasierte Regelungsphilosophie mit dem bekannten LQG-Verfahren zur stochastischen optimalen Regelung linearer Systeme und bietet dadurch eine optimale Möglichkeit, um netzwerkbedingte Datenverluste und Übertragungsverzögerungen zu kompensieren. Die Reglersynthese erfolgt dabei über einen optimierungsbasierten Ansatz. Durch eine geeignete Modellierung der Übertragungscharakteristik der verwendeten Netzwerke kann das resultierende Optimierungsproblem mittels dynamischer Programmierung in *geschlossener Form* gelöst werden.

Ein Vorteil des S-LQG Verfahrens - insbesondere im Vergleich zu anderen optimierungsbasierten Regelungsverfahren wie der modellprädiktiven Regelung (MPC) - ist, dass das S-LQG zur Regelungslaufzeit nur einen geringen Berechnungsaufwand erfordert. Dies resultiert aus der analytischen Lösbarkeit des unterlagerten Optimierungsproblems, da ein Großteil der notwendigen Berechnungen bereits vor Laufzeit der Regelung ausgeführt werden können. Zudem ergibt sich daraus ein sehr geringer Speicherbedarf des Regelungsalgorithmus, sofern Zeitinvarianz des zu regelnden Systems vorausgesetzt werden kann.

In der Arbeit werden ferner wichtige Erweiterungen des Basisverfahrens behandelt. Beispielsweise wird zur Berücksichtigung möglicher Bandbreitenbeschränkungen der Netzwerke eine *ereignisbasierte Betriebsart* des S-LQG vorgestellt. Darüber hinaus wird basierend auf dem S-LQG ein optimales Verfahren zur sequenzbasierten Folgeregelung präsentiert. Im Vergleich zum Stand der Technik zeigen die entwickelten Verfahren in Simulationen eine sehr gute Performanz, sodass die praktische Anwendung in realen Szenarien bereits zusammen mit einem Industriepartner aus dem Bereich der Fertigungsautomation initiiert wurde.

Abstract

Digital data networks have been used in control applications for more than 20 years to connect sensors, actuators, and controllers of a control loop. Typically, highly specialized networks are applied, the so called fieldbuses. If sufficiently dimensioned, these networks ensure deterministic data transmission with guaranteed latency. However, it is often desirable or even necessary to use wireless networks (such as Bluetooth) and/or general computer networks (such as the Internet) instead of the fieldbuses due to higher flexibility, lower costs, and the potential to meet special requirements.

In this way, the Internet can be used within the control loop to bridge long distances as they occur in telerobotic applications, for example. In process and factory automation, wireless networks allow the replacement of costly transmission elements such as slip rings and cable carriers. Also, actuators and sensors can be placed in locations that are hard to access. Moreover, wireless car-to-car communication offers new control potential for intelligent highway systems and self-organizing platooning vehicles.

However, using wireless networks and/or general computer networks within a control loop presents significant challenges since these networks do not have deterministic transmission characteristics. In contrast to the specialized fieldbuses, there is the danger of time-varying transmission delays and stochastic data losses, which are frequently experienced in wireless networks in particular. These network-induced disturbances can massively degrade the control performance and even destabilize the closed-loop system.

In the last 15 years, research has emerged in the area of *Networked Control Systems* (NCS) to investigate these problems at the intersection of communication and control. To date, a plethora of methods have been proposed to consider the network effects during the control design. One of these methods is *sequence-based control* (also referred to as *packet-based control*, *packetized predictive control*, or *receding horizon networked control*). The analysis, extension, and application of the sequence-based method is the main subject of this work.

Sequence-based control uses the property of modern communication networks (such as Ethernet-TCP/IP) that data are sent in the form of packets, which can transport more information than needed for a single control data transmission. A sequence-based controller uses the available capacity of a data packet and not only sends the current control data, but also a sequence of predicted control inputs that can be applied at a future instant. The predicted control inputs can be applied by the actuator if a subsequent transmission gets delayed or lost. In this way, the network-induced effects can effectively be mitigated.

In this work, the newly developed S-LQG (Sequence-Based Linear Quadratic Gaussian) control method is presented. The method combines the idea of the sequence-based control with the LQG approach to stochastic optimal control in order to optimally compensate for network-induced time delays and packet losses. For controller synthesis, the control problem is formulated as an optimization problem that includes a simplified stochastic model of the networks. Using a state augmentation technique, the dynamic programming algorithm can be applied to solve the optimization problem *closed-loop optimal* in *analytic form.*

In comparison to other optimization-based approaches such as MPC (Model Predictive Control), the S-LQG can be calculated offline. This is a great advantage as it also allows the application of the S-LQG in time critical applications due to the low computation requirements during operation. Moreover, assuming a time-invariant plant, the controller gains converge to a steady-state so that the S-LQG only occupies a small amount of memory.

Further, important extensions of the S-LQG are also discussed. For example, an *event-triggered* extension of the proposed approach is presented that can be used in the context of band-limited networks to reduce the required bandwidth. Also an *optimal* tracking controller is derived based on the S-LQG solution that makes optimal use of existing preview information about the reference trajectory. In simulations, the developed approaches show a very good performance compared to state-of-the-art methods such that the application of the S-LQG method in the field of factory automation has already been initiated in conjunction with an industrial partner.

Notation

General Conventions

x	Scalar (lowercase)
$\underline{x}$	Vector (underlined, lowercase)
$\mathbf{A}$	Matrix (bold, uppercase)
$(.)_k$	Quantity at time step k
$(.)^*$	Optimized quantity, result of minimization
$\underline{x}_{0:k}$	Sequence/set of quantities $\{\underline{x}_0, \ldots, \underline{x}_k\}$
$\mathbf{A}(n)$	Matrix depending on integer n
$\mathbf{A}(0:n)$	Sequence/set of matrices $\{\mathbf{A}(0), \mathbf{A}(1), \ldots \mathbf{A}(n)\}$
$\mathcal{A}$	Set (calligraphic, uppercase)
$(.)^{Evt}$	Quantity related to event-triggered approach
$(.)^{Trk}$	Quantity related to tracking control approach
$\mathbf{A} > 0$	Matrix $\mathbf{A}$ is positive definite
$\mathbf{A} \geq 0$	Matrix $\mathbf{A}$ is positive semidefinite
$\mathbf{A} > \mathbf{B}$	Matrix $(\mathbf{A} - \mathbf{B})$ is positive definite
$\mathbf{A}(0:n) > \mathbf{B}$	Abbrevation for $\{\mathbf{A}(0) - \mathbf{B} > 0, \ldots, \mathbf{A}(n) - \mathbf{B} > 0\}$

Operators

$\mathbf{A}^\top$	Matrix transpose of $\mathbf{A}$
$\mathbf{A}^\dagger$	Moore-Penrose pseudoinverse of $\mathbf{A}$
$\mathbf{A}^{-1}$	Inverse of $\mathbf{A}$
$\mathrm{eig}\,(\mathbf{A})$	Set of eigenvalues of $\mathbf{A}$
$\mathrm{tr}\,(\mathbf{A})$	Trace of $\mathbf{A}$
$\delta_{(x,y)}$	Kronecker delta function
$\mathrm{Prob}\,(a)$	Probability of event a
$\mathrm{Prob}\,(a\|\, b)$	Probability of event a given event b
$\mathrm{E}\,\{x\}$	Expected value of x
$\mathrm{E}\,\{x\|\, y\}$	Expected value of x conditioned on y
$\mathcal{A} \cup \mathcal{B}$	Union of $\mathcal{A}$ in $\mathcal{B}$
$\mathcal{A} \backslash \mathcal{B}$	Relative complement of $\mathcal{B}$ in $\mathcal{A}$

Symbols

$\mathbf{I}$	Identity matrix
$\underline{0}$	Vector where all entries are zero
$\mathbf{0}$	Matrix where all entries are zero
$\mathbb{N}_0$	Set of natural number including zero
$\mathbb{N}_{>0}$	Set of natural number excluding zero
$\mathbb{R}$	Set of real number
$\mathbb{R}^n$	n-dimensional vector space over the field of the real numbers
$\emptyset$	Empty set
$\square$	End of proof or lemma

Conventions for Variables

$\underline{x}_k$	System state at time k
$\underline{\xi}_k$	Augmented system state at time k
$\overline{\underline{x}}_0$	Initial system state
$\mathbf{\Lambda}_0$	Covariance of the initial system state
$\underline{y}_k$	Measurement obtained by sensor at time k
$\mathcal{Y}_k$	Set of measurements received by controller at time k
$\underline{w}_k$	Process noise at time k
$\underline{v}_k$	Measurement noise at time k
$\underline{z}_k$	Plant output at time k considered for tracking task
$\underline{z}_k^{Ref}$	Reference value at time k
$\underline{u}_k$	Control input applied by the actuator at time k
$\underline{U}_k$	Sequence of control inputs calculated at time k
$\underline{u}_{k\|m}$	Control input calculated at time m, applicable at time k
N	Length of control sequence $\underline{U}_k$, capacity of actuator buffer
$\underline{u}_k^{df}$	Default control input applied by actuator if buffer is empty
$\mathbf{A}_k$	System matrix
$\mathbf{B}_k$	Input matrix
$\mathbf{C}_k$	Output matrix
$\mathbf{W}_k$	Process noise covariance
$\mathbf{V}_k$	Measurement noise covariance
$\mathbf{Z}_k$	Output matrix considered for tracking control

$\mathbf{Q}_k$	Weighting matrix of cost function for system states
$\mathbf{R}_k$	Weighting matrix of cost function for control inputs
$\mathcal{I}_k$	Information set available to the controller at time k
K	Terminal time of control task
$C_{0 \to k}$	Expected cumulative costs from initial time to time k
$J^*_{k:K}$	Minimum expected *cost-to-go* from time k to terminal time
θ_k	State of Markov chain describing the age of the control sequence buffered in the actuator at time k
$\mathbb{J}$	Co-domain of θ_k, subset of $\mathbb{N}_0$
$\mathbf{T}$	Transition matrix of θ_k
$p(j,i)$	Entry in j-th row and i-th column of $\mathbf{T}$
τ_k^{CA}	Delay in controller-actuator network at time k
τ_k^{SC}	Delay in sensor-controller network at time k
$q^{CA}(m)$	Probability that $\tau_k^{CA} = \mathrm{m}$
$q^{SC}(m)$	Probability that $\tau_k^{CA} = \mathrm{m}$
μ_k	Control law at time k
$\mathbf{L}_k$	Controller gain matrix at time k

Glossary

DDE	Delayed Differential Equation
LMI	Linear Matrix Inequality
LQ	Linear Quadratic
LQG	Linear Quadratic Gaussian
LQR	Linear Quadratic Regulator
MAC	Media Access Control
MJLS	Markov Jump Linear System
MMSE	Minimum Mean Square Error
MPC	Model Predictive Control
NCS	Networked Control Systems
S-LQG	Sequence-Based Linear Quadratic Gaussian
SVD	Singular Value Decomposition
TCP	Transmission Control Protocol
TDS	Time Delay System
UDP	User Datagram Protocol
i.i.d.	Independent and identically distributed

1. Introduction

Feedback is a fascinating phenomenon that always occurs if the output of a system is fed back to its input. The resulting interplay between output and input blurs the line of causality since the output not only becomes effect of the input but, at the same time, also is the input's cause. In this way, feedback can change the characteristics of the original system beyond recognition and may create something new. It may even be the source of human consciousness as Hofstadter postulates in his popular book "Gödel, Escher, Bach" [51]. In this thesis, we cannot go this far but we will use feedback as an extremely powerful engineering tool to control things and make them do what we want them to.

In a recent publication, Åström and Kumar describe the long history of feedback control in engineering and trace its developments from the Industrial Revolution to now [8]. They identify that control was entering a new stage around the year 2000. Traditional applications recorded a significant spread due to falling computing costs, and new fields of control emerged. Major drivers included the increasing desire for distributed and autonomously operating systems as well as the technological advances in computing and communication.

In particular, enhanced reliability of wireless communication systems, growing availability of cost effective general computer networks, and the global expansion of the Internet laid the foundations for a new era of networked applications [87]. For example, Internet-based teleoperation and long distance remote control were explored, and new concepts for automated wireless highway and traffic systems as well as unmanned aerial vehicles (UAV's) were conceived. Moreover, wireless applications became one of the leading technological trends in factory and process automation [4].

However, using wireless networks and general data networks within a feedback loop confronted control engineers with new challenges. In contrast to the specialized, wired fieldbuses used so far, these new networks do not provide deterministic transmission characteristics. It was recognized that further research is necessary to face the resulting challenges [87] and the research area of *Networked Control Systems* (NCS) emerged. In this context, new analysis tools and control methods have been developed to deal with the network induced effects.

One of these new methods is *sequence-based control* [107], which is (among others) also known as *packet-based control* [167], *packetized predictive control* [113], and *receding horizon networked control* [45]. The method is specifically designed to compensate for transmission losses and time varying network delays that occur in wireless networks or the Internet. The method exploits the property of these communication networks that data are sent in the form of data packets, which can usually transport a higher data load than needed for a single control data transmission. The idea is to utilize this spare data capacity by sending not only the current control data, but also information that can be beneficially used by the receiver to compensate for subsequent packet losses or transmission time delays.

Based on the sequence-based approach, this work presents the newly developed Sequence-based Linear Quadratic Gaussian (S-LQG) control method (own publs. [170–176]). The S-LQG combines the sequence-based control philosophy, that compensates for network-induced transmission delays and losses, with the well-known LQG (Linear Quadratic Gaussian) theory of stochastic optimal control. The problem setup under consideration and the contributions of this work are briefly summarized in the next section.

1.1. Problem Formulation and Contribution

In this work, we deal with the problem of controller design for centralized NCS that are subject to stochastic transmission losses and time-varying packet delays. The basic setup is schematically depicted in Fig. 1.1. It can be seen that the controller is connected to the actuator and the sensor via

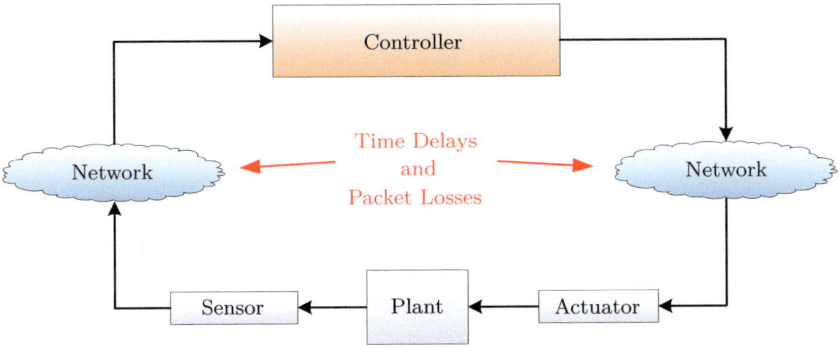

Figure 1.1.: General NCS setup under consideration

data networks. The data networks introduce various disturbing effects into the control loop and impose constraints not considered in standard control theory. However, to avoid tremendous degradation of control performance, the network effects have to be considered in the controller design. In this thesis, we are concerned with two of the most severe network effects: stochastic transmission losses and time-varying packet delays. Our aim is to design a controller that is able to optimally compensate for these network-induced disturbances.

To that end, we extend the sequence-based control approach (see e.g. [70]). In sequence-based control, the controller not only computes a single control input per time step, but also a whole sequence of predicted inputs applicable at future time steps. These control inputs are all sent within the same data packet to the actuator. In case a subsequent packet transmission gets delayed or lost, the actuator can fall back upon the predicted control inputs to apply reasonable inputs to the plant.

Our approach to sequence-based control is from stochastic optimal control and, in particular, LQG theory. The main contribution of this work is the unification of LQG theory with the sequence-based control method. Specifically, if one can assume linear system dynamics and a so called TCP-like (Transmission Control Protocol) network protocol providing idealized acknowledgments between controller and actuator [55], we are able to derive the *optimal* sequence-based controller in the LQG sense (own publ. [174]). Therefore, we also refer to this optimal controller as S-LQG (Sequence-Based Linear Quadratic Gaussian). Based on the stochastic transmission characteristics of the networks, the S-LQG optimally accounts for both transmission losses and stochastic delays. This is in contrast to former work, where only losses were considered [45].

An advantage of the S-LQG is that the control law can be computed offline. Therefore, compared to other optimization-based approaches such as MPC (Model Predictive Control), the computing resources required during operation are relatively small and allow for time critical applications such as motion control. In addition, the S-LQG has only a small demand for memory space even for long time horizons due to convergence of the time-varying feedback gains. Furthermore, the expected performance of the S-LQG, measured in terms of a quadratic cost function, can be expressed in closed-form. This allows us to derive new stability criteria for linear sequence-based controllers with quadratic costs (own publ. [172]).

The closed-form solution also serves as the basis for an *event-triggered* extension of the S-LQG that reduces the network load to consider possible band-limitations (own publ. [170]). In addition, an optimal *sequence-based tracking controller* will be presented. Based on the S-LQG results, this controller optimally incorporates set point changes and available preview information on the reference trajectory (own publ. [173]). Moreover, a practical extension of the results will be presented, allowing for generalization to network protocols that do not provide acknowledgment signals such as UDP/IP (User Datagram Protocol/Internet Protocol) networks (own publ. [175]).

Finally, we also mention what is not covered in this work. As we concentrate on centralized optimal control solutions in a linear setup, we do not touch the topic of decentralized optimization and multi-agent systems. Also, the extension to nonlinear systems and the incorporation of quantization effects and imperfect clock synchronization has been left to future research. Finally, we are only concerned with the details of specific communication technologies, protocol architectures, routing algorithms, MAC-layer designs, etc. in so far as they are necessary to describe the effects perceived within the feedback loop from a control related perspective.

1.2. Outline of the Thesis

Chapter 2 provides a general introduction to NCS and its emergence as one of the most active research areas within the control community. The specific challenges arising in NCS are summarized in Section 2.2. State-of-the-art control methods addressing these challenges are introduced in Section 2.3, and the sequence-based method is described separately in Section 2.4.

The main result of this work is presented in Chapter 3. It comprises the detailed derivation and the discussion of the optimal sequence-based LQG controller (S-LQG). To that end, the sequence-based NCS system setup is introduced in Section 3.1, and the optimal control problem formulated in Section 3.2. Then, in Section 3.3, a sequence-based NCS model is derived that captures the controller-actuator network and the sequence-based buffering scheme of the actuator. Based on this, the optimal S-LQG controller is presented and evaluated in Section 3.4. Furthermore, a stability analysis is carried out in Section 3.5 and minor extensions of the algorithm are presented in Section 3.6.

The S-LQG stabilizes the plant around the zero state. In Chapter 4, an optimal sequence-based tracking controller is presented that not only addresses non-zero set points but to follow given reference trajectories. The NCS setup and the problem formulation are described in Section 4.1. The optimal solution is discussed in Section 4.2. and the resulting controller demonstrated in simulations in Section 4.3.

The S-LQG compensates for time delays and packet losses by sending additional information with the data packets. The additional information can increase the network load such that it has a negative impact on the control performance. In order to avoid this, and to address possible band-limitations, an event-triggered operation mode of the S-LQG is presented in Chapter 5. The problem is formulated in Section 5.1 and the solution derived in Section 5.2. The stability of the proposed event-triggered controller is investigated in Section 5.3 and the performance evaluated by simulations in Section 5.4.

Finally, this work concludes with a summary of the results obtained and an outlook on possible future research topics in Chapter 6.

2. Networked Control Systems

Looking for a broad definition of the term *Networked Control Systems (NCS)*, we quickly find NCS described as "closed-loop systems that have to be considered as networked systems" in the recent publication [77]. This definition comprises two aspects. First, NCS are characterized as feedback control systems that use data networks and, second, the data networks impose challenges that cannot be considered independently of the feedback control system. Interestingly, although networks have been used in feedback control systems since the early 1980s, the term *Networked Control Systems* did not appear before the late 1990s [61, 144]. This reveals a shift in the perception of networks in feedback control with NCS becoming one of the most active research areas within the control community today. In the next section, we briefly trace the emergence of NCS on a general level. Then, the specific challenges experienced in NCS are explained in Section 2.2, and state-of-the-art control methods to counteract these problems are reviewed in Section 2.3. In particular, we give special emphasis to the sequence-based control method that is described separately in Section 2.4.

2.1. Emergence and Overview of the Field

Digital data networks were introduced to the area of feedback control about 30 years ago to connect spatially distributed components of a control loop. In contrast to the multiple wire topologies used up to that time, digital networks only required a single data line to connect the control components and therefore reduced installation costs. Single wire solutions, however, present a bottleneck in the system with regard to the possible speed of data transmissions. To satisfy the demanding real-time requirements of industrial applications, highly specialized networks have been developed that are jointly referred to as *fieldbuses*. One of the first fieldbuses on the market was CAN (Controller Area Network), which was developed by the German company Robert Bosch in 1983 [20]. Other fieldbuses followed, such as Profibus introduced in 1987 [106], Interbus and DeviceNet developed around same time [101], and Foundation Fieldbus initiated in 1994 [32]. Sufficiently dimensioned, fieldbuses ensure deterministic data transmissions with guaranteed latency. We therefore refer to this kinds of networks as *deterministic networks*. They provide proprietary communication solutions for control applications and are standard in industrial control today.

Around the year 2000, efforts increased to utilize networks with *non-deterministic* transmission behavior for control purposes [86]. The desire was fueled by two technological developments. These are the remarkable improvements of wireless communication technologies [8, 130] on the one hand, and the advances in general computer networks and, in particular Ethernet, on the other hand, that also paved the way for the expansion of the Internet.

In particular, general computer networks recorded falling prices, increased speed, and gained expanded interoperability and reliability. As a result, networks such as Ethernet-based LAN were brought to various institutions and companies and a widespread infrastructure has been formed. At the latest with the connection to the Internet, intense endeavors have been undertaken to employ these networks for feedback control as well [86]. Relevant applications include Internet-based factory and building automation, as well as remote process control [161]. Long distance telerobotics [6, 21, 150], telesurgery [80], and haptics collaboration [50] have also been pursued. The advantage of general computer networks, such as the Internet, is that the already existing communication infrastructure can be used without additional installation costs. In contrast to fieldbuses, these networks are also non-proprietary, so no licensing fees have to be paid.

Wireless network technologies such as Bluetooth (IEEE 802.15.1), WLAN (IEEE 802.11), and ZigBee (based on IEEE 802.15.4) have experienced fast progress over the most recent ten years and have become strong innovation drivers for control applications [57, 104]. For example, wireless networks broaden the scope of feedback control to mobile objects such as unmanned aerial vehicles (UAV) or ambient intelligent networks [68], and allow the placing of actuators and sensors in locations that are hard to access. Furthermore, they can be used to replace mechanically stressed elements such as slip rings or cable carriers as found in electric monorail systems, automatic storage systems, and gantry cranes. In this context, wireless networks not only reduce maintenance and installation costs, but also increase the availability of the system [18]. The emerging field of wireless car-to-car communication also offers new control perspectives for automated highway systems, platooning vehicles, traffic supervision, and intelligent cross-road-management [121, 122].

However, using general data networks and/or wireless networks in feedback control systems also creates significant challenges. In contrast to deterministic fieldbuses, these networks generally have *non-deterministic* transmission characteristics. This means that not only can transmission losses occur, as observed in wireless networks in particular, but also that data transmissions may be subject to unknown and time-varying delays. Furthermore, if the network is shared with other applications, several sources compete for network access resulting in additional restrictions for the control system. Experience has shown that the network-induced effects can have a substantial negative impact on the performance of a feedback system. For instance, even small variations of transmission times can destabilize a control system [163].

Therefore, the need was recognized for new control strategies and analysis tools that are able to deal with the non-deterministic network effects. Research in this direction commenced at the late 1990s and soon became known as Networked Control Systems. A characteristic of NCS research is the strong interaction of control engineers with computer scientists and communication engineers [77]. The common objective is the combination of control theory, which describes dynamical systems connected by perfect links, with the area of communication theory, which addresses information transported through imperfect links. Further information on the general field of NCS and the relation to other active research areas such as *Cyber-Physical Systems* can be found in the overviews [4, 9, 12, 66, 77, 142].

In the following section, we discuss the specific challenges introduced by the non-deterministic networks in more detail.

2.2. Challenges in NCS

In this section, we describe the main challenges arising when non-deterministic networks are integrated into a feedback loop. The challenges are categorized from the perspective of feedback control, i.e., grouped regarding their appearance in the control loop. This means that we only go into the details of communication technology and protocol layers where this is necessary and do not strictly separate effects caused by single point-to-point transmissions from effects induced by the interaction of several network nodes.

2.2.1. Time Delays

A main source of performance degradation in NCS originates from the presence of unknown or time-varying transmission delays. As shown in [163], even comparably small variations in the time delays may lead to instability of the entire feedback system. In NCS, different sources of time delays can be identified [65, 66, 161]. On the one hand, these are the computational delays of the digital control components and of the controller in particular. On the other hand, we are faced with network-induced delays resulting from communication related computations at sending and receiving nodes (e.g. for coding and decoding the data), waiting times to gain network access, and transmission delays due to the transportation and routing of the data through the medium[1]. In many NCS scenarios, the network-induced delays represent the critical component of the overall time delay. Computational delays are usually of minor importance and indeed are often negligible [161]. The network access and transportation delays can be very volatile due to varying network conditions (e.g. changing routing paths) or unsteady data load. In regard to the control of NCS, time-varying delays are much harder to compensate for than constant time delays. Therefore, we distinguish between the two in the following sections.

[1] In this work, the terms *transmission delay* and *time delay* are used synonymously.

Time-Invariant Delays: Constant delays occur in NCS that use (sufficiently dimensioned) fieldbuses. Fieldbuses use specific hardware mechanisms to accurately synchronize all distributed communication components so that a sufficiently small jitter can be ensured [123]. Data collisions are either avoided by using a deterministic protocol to manage network access (e.g., in ControlNet and Profibus [123], EtherCAT [29], or FlexRay [98]) or data collisions are resolved by prioritizing the data packets (such as in CAN bus [20]). In the latter strategy, however, the fieldbus must be dimensioned to meet the maximum possible network load [65]. Due to the omnipresence of dead time in all kind of applications, control of systems with constant delays has been intensively investigated for more than 50 years. Theory has reached a mature state and well-known methods are available such as the Padé-approximation or the Smith predictor [76, 128]. In this work, constant delays are therefore of minor interest and only seen as a special case of time-varying delays.

Time-Varying Delays: If networks other than fieldbuses are used for real-time control purposes, we generally have to expect that any time delays that occur will be of a time-varying nature. In particular, this is true for general computer networks such as Ethernet and the Internet [130]. Ethernet, for example, uses the CSMA/CD (Carrier Sense Multiple Access/Collision Detection) method for medium access control that introduces a random delay time if a collision is detected. Time-dependent network load, different routing paths, and various data queues in the network components are further sources for time-varying delays [100]. In wireless networks, the transmission characteristics also strongly depend on the surrounding environment [57]. Mobile objects and obstacles in the line of sight, interferences with external sources, or fading due to reflections and bending of electromagnetic waves can cause a time-varying SNR (signal to noise ratio) of the received signal. As a result, time-varying delays occur due to retransmissions of lost data packets, resynchronization, or restoration of lost connectivity.

In Section 2.2, we will review control strategies for NCS with time-varying delays. These methods can be categorized in terms of the model used to describe the time delays. We therefore distinguish whether time-varying delays are non-deterministically varying [163], or stochastically varying [151].

- *Non-deterministically Varying Time Delays*: In the non-deterministic model, it is assumed that time delays are time-varying and not known in advance. Information is only available on the basis of ensemble characteristics of the time delays such as minimum and/or maximum value.
- *Stochastically Varying Time Delays*: In the stochastic time delay model, time delays are also time-varying and unknown in advance. However, information is available on the probability that certain time delays will occur. The information is usually given in the form of probability density functions over the time delay distributions of the networks. In this work, we assume that a stochastic time delay model is available.

As individual data packets may suffer different time delays, it is possible for data packets to be received in a different order than the sending order. This is known as *packet disordering* and has to be detected on the receiver side to avoid misinterpretation of the data [161]. In this work, we assume that data packets are marked with an index or time stamp so that packet disordering can be easily resolved.

2.2.2. Packet Losses

In contrast to time delays, packet losses have not been intensively investigated in classic control theory [77]. In NCS, packet losses occur for many reasons [130]. One of the most common sources in general computer networks is congestion due to high network load. Components of a network infrastructure such as repeaters and switches can only hold a finite number of data packets in their internal queues. If this memory is fully occupied, additional incoming data packets are discarded. Packet losses can also be caused by transmission failures resulting from malfunctions of network components or unsuccessful decoding at the receiver node. In particular, wireless networks are extremely sensitive to environment-induced disturbances such as interferences and fading. This often makes successful decoding impossible, despite error correction methods such as FEC (forward error correction) or diversity techniques [57]. Indeed, the average packet loss rate of a wireless channel in a typical industrial environment can reach 60% [147].

Packet losses and time delays are closely related and can often be converted back and forth. For example, using a TCP/IP protocol, data packets will be automatically retransmitted if not received within a specified time [130]. This prevents data losses by transforming them into time delays, but potentially increases the average packet loss rate due to the higher network traffic. On the other hand, in many control applications, it is not useful to process measurements and control data that have been delayed for a long time [57]. Therefore, packets not received within a certain time are usually discarded and, hence, time delays are transformed into packet losses.

2.2.3. Band-Limitations

Further challenges arise when the amount of data that can be transported through the network [161] is constrained. These constraints may be directly imposed by bandwidth limitations of the data channels in use, or also by the control application itself. Examples are UAV's (Unmanned Aerial Vehicles) in stealth mode that only use minimum communication to avoid detection or applications with energy constraints such as underwater vehicles or planetary rovers. In compliance with [49], we summarize this kind of network-induced challenge under the term *band-limitations*.

In point-to-point connections, band-limitations can be investigated from a viewpoint of *data quantization* as the problems encountered are very similar to the case of analog-digital conversion with finite word length [89]. In this context, one aim is to find optimal coding/decoding-schemes satisfying given data rate constraints [159]. Another line of research focuses on band-limitations resulting from the presence and organization of multiple network nodes. In these scenarios only a subset of the connected nodes can simultaneously be granted access to the network. As a result, *competing network accesses* occur and must be resolved [144]. Generally, this is accompanied by additional time delays and packet losses so that the problem of band-limitations is also closely related to the problems discussed in the previous sections 2.2.1 and 2.2.2.

2.2.4. Other Challenges

For the sake of completeness, we will briefly state further challenges occurring in NCS that, however, are not directly addressed in this work.

Clock Synchronization: Due to the spatial distribution of the NCS components, the different clocks have to be synchronized over the data network. The occurrence of time-varying delays and packet losses in the network complicates accurate synchronization [161]. The problem has received much attention, not only in the context of real-time capable fieldbus technologies, but also in general computer networks so that a plethora of methods have been proposed to deal with these effects [56].

Time-Varying Sampling Intervals: In standard control theory, data are assumed to be sampled periodically. In NCS, however, this might no longer be the case [133]. In particular, band-limited networks or inaccurate clock synchronization may lead to non-periodic sampling times. In the NCS literature, this problem is also referred to as *time-varying transmission intervals* and has been investigated in the context of *event-based* control in particular (see Chapter 5).

Security: NCS are often integrated into network structures that also provide other data services. As a result, there may be many access points to the network, in particular, if the NCS is connected to the Internet. This opens the door for hacking attacks and sabotage as demonstrated by the computer virus Stuxnet [63]. Therefore, security mechanisms have been developed that semantically analyze any control commands given from a control perspective to detect possible intruders [25].

2.3. Control Methods

To address the challenges described in the previous section, a plethora of control methods have been proposed in the literature. General overviews can be found in [49, 69, 138, 153, 161], for example. In [138], the state-of-the-art through the year 2003 is presented. This is extended in [49] and [161] by the methods developed from 2003 to 2007, and from 2007 to 2013, respectively.

In the following summary of NCS control methods, we concentrate on approaches that are mainly concerned with the stabilization problem in the presence of time-varying transmission delays and packet losses. Approaches to NCS tracking control will be treated in Chapter 4 and approaches dealing with band-limitations will be described in Chapter 5 in the context of the proposed *event-triggered* method. Here, we divide the approaches into two classes. The first class of these methods only uses a-priori information about the network effects. This means that the controller does not consider any information about currently occurring time delays or packet losses during run time. We therefore refer to these approaches as *non-adaptive* methods. In contrast, the second class of methods is characterized by its ability to incorporate online measurements of time delays and packet losses into the control design. By doing so, the controller adapts to the current network situation. These approaches are therefore labeled as *adaptive* methods. The sequence-based control method belongs to the class of adaptive methods and, due to its importance, is discussed separately in Section 2.4.

2.3.1. Non-Adaptive Methods

In this section, we discuss control methods that only use offline information on time delays and packet losses that occur. Therefore, the approaches described each have in common that the resulting control law is independent of the realizations of time delays and packet losses experienced during run time. Hence, the controller must be robust against all possible realizations of network-induced effects and therefore is often optimized for the worst case scenario. This has the advantage that time delays and packet losses do not have to be detected online. On the downside, this leads to control designs that are more conservative than adaptive methods which use online information about the current network situation.

Time-Varying Delays: One of the earliest non-adaptive methods to compensate for time-varying delays is the so called *Queuing Method* [75]. In this approach, queues are used at the input of controllers and actuators to buffer incoming data packets. A data packet is released from the buffer once a specified time, the so called playback time, has elapsed since creation of that data packet. In this way, time-varying delays are transformed to constant delays (at the expense of additional time delays) so that well-known control methods for constant delays can be used. A recent adaption of this method in combination with Smith Predictors and PI controllers can be found in [1], where the queues are also referred to as *playback buffers*. Assuming bounded time delays, the *Lyapunov-based* approach described in [163] derives a stability criterion in terms of an LMI (Linear Matrix Inequality) feasibility problem. The criterion is not only applicable for periodic sampling, but also for time-varying sampling times. A good example to consider concerning time-varying delays in the frequency domain is the robust approach presented in [39]. Time delays are assumed to be bounded and approximated by Padé-all-passes such that network effects can be incorporated as multiplicative noise within a robust H_∞-control design. Furthermore, *fuzzy control* [169] has been applied to NCS for delay compensation, as well as stochastic optimal control [46, 67]. Finally, it is mentioned the important class of controllers that are based on the *Markov Jump Linear System (MJLS)* approach [141]. One of the first methods using a MJLS model in NCS is the work [151] where an output feedback controller is derived.

Packet Losses: A prominent non-adaptive approach for compensating for packet losses in the absence of time delays is summarized in [163]. In this approach, packet losses are considered on the basis of the average loss rate and are incorporated into the NCS model by a switch that closes the control loop at a given rate. The slowest possible average transfer rate that stabilizes the system can then be calculated using *Asynchronous Dynamical Systems (ADSs)* theory [47]. *Fading networks* with packet losses are discussed in [26]. The network is separated into a deterministic part and a stochastic zero-mean multiplicative perturbation. A stabilizing *robust controller* is then derived by solving a non-convex structured minimization problem via D-K-iteration. Another robust approach that only uses offline information of packet losses is described in [146]. The data losses are modeled as stochastic Bernoulli variables which are then replaced by their expected values in order to derive a stabilizing controller with guaranteed satisfaction of a prescribed H_∞-performance level. Moreover, *MJLS* approaches have also been applied for packet loss compensation. A example is the work [149], where correlated losses are modeled independently in both networks by two separate Markov chains.

Time Delays and Packet Losses: The former approaches are applicable to either time-varying transmission delays or packet losses. A typical *Lyapunov-based* approach that is able to consider both effects simultaneously is described in [158]. Again, this approach requires that time delays and consecutive packet losses are bounded. The resulting controller ensures stability and can be computed by the solution of an LMI. An extension of this approach to the problem of reference tracking is discussed in [34]. In addition, the non-deterministic time-varying approach of [163] can be extended to consider both effects simultaneously as shown in [117]. Another line of research in this category originates from the area of *Time-Delay Systems (TDS)* [116]. In this methodology, the NCS is modeled as a *Delayed Differential Equation (DDE)* for which stability results are derived by finding a Lyapunov-Krasovskii functional or application of the Razumikhin theorem [23, 88, 157].

2.3.2. Adaptive Methods

In contrast to the approaches in the previous section, we now survey methods that use online information about time delays and packet losses to adapt to the corresponding situation. Hence, the main characteristic of these approaches is that the applied control inputs depend on the actual time delays and/or packet losses experienced during control operation. It is important to note that although these controllers adapt to the network effects, the corresponding control laws can be calculated offline (with the exception of the Model Predictive Control (MPC) approaches discussed in Section 2.4.3). As the adaptive control methods use more information than the non-adaptive ones, the former have the potential to provide better performance results. However, this comes at the expense of more complex controller structures and usually requires the actuator to be capable of performing minor computations. Also, time delays must be measured. To that end, data packets are often marked with timestamps. This allows the reconstruction of the time delays as long as the components of the control loop are sufficiently synchronized.

Time-Varying Delays: The stochastic LQG method developed in [90, 93] is one of the first approaches that uses controller gains that adapt to the actual time delays experienced. The computation of the controller requires solving a Riccati equation involving expectations with respect to the network effects. The approach has been extended to non-periodic sampling in [92] where possible clock drifts are also considered. In this context, the separation principle known from standard LQG control still holds in the setup under consideration [93]. The original approach is only applicable for time delays smaller than the sampling period. Yet, it is extended in [125] to cover longer delays. Other important approaches that incorporate online measurements of time delays are the stochastically switched MJLS system approaches ([124, 151, 162], see [16] for a general overview). In the early work [151], delays between sensor and controller are modeled by a Markov chain, and the controller adjusts according to the actual delay occurring. The approach has been extended in [162] to incorporate delays between the controller and the actuator where it is assumed that the time delay is known at the controller site within the same time step. In [124], a variation of this approach is presented in which the current delay of the control packets is measured at the actuator site and then sent over the sensor-controller channel.

Packet Losses: To compensate for packet losses, a stochastic approach based on a MJLS model is described in [121, 122]. The controller is located at the actuator site and the feedback gain is adapted when a measurement gets lost in the network between the sensor and the controller. The control law is formulated by means of an LMI and guarantees stability in the H_∞-sense. Using a non-deterministic loss model of the network, the switched system approach to compensate for delays [163] can also be used for the compensation of packet losses [156]. Here, the controller must also be collocated with the actuator.

Time Delays and Packet Losses: Approaches that are able to adapt to both time-varying delays and packet losses almost exclusively belong to the class of *sequence-based control* methods. Due to the high relevance of the method in the context of this work, we will discuss the sequence-based method separately in Section 2.4. Among the rare non-sequence-based approaches, we mention here the approach described in [137] that relies on the queuing method to compensate for delays. The authors use a second buffer to store the history of applied control inputs at the actuator. If a packet loss occurs, the applied control input is calculated by interpolating the buffered history.

2.4. Sequence-Based Control

Sequence-based control is a network-adaptive control method used in NCS to compensate for network-induced transmission delays and data losses between controllers and actuators. The method is also referred to as *packet-based control* [167], *packetized predictive control* [113], and *receding horizon networked control* [45]. In this work, we consistently use the term *sequence-based control* to emphasize that sequences of control inputs are sent over the network instead of single control inputs. The method was first presented in [11] in the context of MPC. Since then, a plethora of variations have been proposed.

2.4.1. The General Idea

Sequence-based controllers exploit the property of modern communication networks that data are transmitted in atomic data packets. The data payload of these packets is usually much higher than the size of a single control input. Thus, the idea is to use the unoccupied payload of the packet by transmitting not only the current control input to the actuator but also a sequence of control inputs that are applicable at future time instants. The additional inputs are buffered at the actuator and can be applied in case a subsequent data transmission gets delayed or lost. In this way, the effects of time-varying transmission delays and packet losses are mitigated.

However, the additional control inputs increase the network load and, thus, can also cause additional time delays and/or packet losses. To reduce this effect, long control sequences can be parameterized by appropriate compression techniques [54] or an event-triggered strategy can be used [37, 168]. The latter strategy will be discussed in Chapter 5. Even without these techniques, the performance increase due to the additional control information typically outweighs the deterioration of the network quality as demonstrated by many experimental studies with various networks [72, 94,134,136,140].

The sequence-based method is extremely powerful in combination with networks that have a fixed minimum payload size such as Ethernet (IEEE 802.3). An Ethernet packet contains a 112-bit or 176-bit header and at least 368-bit of payload [100]. If less than 368-bit of data shall be transmitted, the data frame is filled with zeros. Therefore, using this data space does not degrade the network quality at all. The same holds for networks that use the so called Asynchronous Transfer Mode (ATM) protocol. In ATM, the data packets are referred to as cells and have a fixed size with 40-bit of header information and 384-bit for data [130].

To implement a sequence-based control strategy, the actuator must have enough computational resources to process incoming control sequences and apply appropriate selection logic. Typically, these computing resources already exist as the actuator uses a digital communication interface to communicate with the controller. Another requirement for implementing a sequence-based strategy is that the actuator is able to apply the predicted control inputs at the intended time instances. Therefore, the actuator has to work in a time-driven mode, which also requires that the internal clocks of the controller and the actuator are sufficiently synchronized. Methods for obtaining synchronization are described in [56]. However, there are also sequence-based methods that overcome this synchronization procedure (at the expense of more communication).

In the following, we discuss the state-of-the-art approaches to sequence-based control. The methods are grouped in three classes according to the applied principle of control sequence generation. We distinguish methods that are based on a nominal controller (Section 2.4.2), sequence-based MPC methods (Section 2.4.3), and stochastic optimal control approaches (Section 2.4.4).

2.4.2. Methods Based on a Nominal Controller

The most popular approach to sequence-based control is via the extension of a so called nominal controller. The nominal controller is not sequence-based and is usually designed without taking the network-induced effects into account. Control sequences are then calculated by predicting the future outputs of this nominal controller over a finite horizon based on predictions of the future states. The main advantage of the approach is that the nominal controller can be designed by any standard control method such as PID parameter tuning, pole placement, H_2-, or H_∞-methods. In particular, the approach is very convenient if a nominal controller has already been designed for the non-networked system.

To illustrate the idea of this approach, consider that a nominal controller has already been designed and is described by the state feedback matrix $\mathbf{L}$ such that

$$\underline{u}_k = \mathbf{L}\underline{x}_k \ , \tag{2.1}$$

where $\underline{u}_k$ and $\underline{x}_k$ are the control input and the current system state, respectively. Then, a control sequence U_k of length N is calculated at time step k by

$$\underline{U}_k \stackrel{\text{def}}{=} \begin{bmatrix} \underline{u}_{k|k} \\ \underline{u}_{k+1|k} \\ \underline{u}_{k+2|k} \\ \vdots \\ \underline{u}_{k+N-1|k} \end{bmatrix} = \begin{bmatrix} \mathbf{L}\,\underline{x}_k \\ \mathbf{L}\,\hat{\underline{x}}_{k+1} \\ \mathbf{L}\,\hat{\underline{x}}_{k+2} \\ \vdots \\ \mathbf{L}\,\hat{\underline{x}}_{k+N-1} \end{bmatrix} , \tag{2.2}$$

where $\underline{u}_{m|k}$ are control inputs calculated at time k and intended to be applied at time m, and $\hat{\underline{x}}_{k+1}, \dots, \hat{\underline{x}}_{k+N-1}$ denote the predictions of the future states $\underline{x}_{k+1}, \dots, \underline{x}_{k+N-1}$.

Dependent on the system setup and assumptions made, several variations of this basic scheme have been proposed in literature.

State-of-the-Art

The approaches [60, 73, 94] use the basic scheme (2.2) in combination with a non-deterministic network model to predict the states. Assuming that the system is linear and deterministic and that time delays and consecutive packet losses are bounded, conditions for closed-loop stability have been derived in [73] using a switched systems approach. In this case, the conditions explicitly depend on the given controller and observer gains. A possible extension to incorporate model uncertainties and avoid clock synchronization in practical applications is presented in [53]. The case of possibly unbounded consecutive packet losses has also been considered [94]. In this approach, data packets carry the information of whether the corresponding control sequence has been calculated based on a current measurement or based on a state estimate derived from older measurements. Then, if a sequence does not arrive at the actuator in time, the buffered sequence is used until a new control packet is received that is based on a current measurement. The approach is extended in [140] to wireless networks where experimental results are also shown.

Another group of approaches combines the idea of sequence-based control with the *queuing method* described in Section 2.3.1 [22,30,41,44,70,105,132]. The approach in [70] can be interpreted as an extension of the queuing method where not only the current control input is transmitted to the actuator but also the (possibly already) queued content of the playback buffer is permanently retransmitted. This has the advantage that only the last control input of the control sequence has to be computed at each time step as the other inputs have already been calculated in previous time steps. Also, assuming an undisturbed deterministic linear system and bounded network effects, the closed-loop stability of the system is independent of the network if the control sequences are long enough. The approach is extended in [132] to also incorporate possible disturbances. Another very popular approach in this category is described in [105]. Therein, the queuing method is not used on the level of single control inputs, but applied to entire control sequences. This means that control sequences are calculated to be applicable beginning at a certain future time step. If such a sequence is received by the actuator, it is first buffered and only activated when the specified playback time is reached. In the case that a sequence has not been received in time, additional data are sent via the sensor-controller channel to avoid inconsistencies. The approach is applicable to nonlinear noiseless plants with model approximation mismatch and does not require clock synchronization. It has been further extended in [41] and [30] to also consider multiple sensors and dynamic controllers, respectively.

In contrast to the sequence-based controllers described so far, the approaches [59,107,154] use stochastic network models for time delays and/or packet losses. In [59], a stabilizing sequence-based controller with H_∞ disturbance attenuation is derived where data losses are modeled with a so called *Gilbert Elliott Model.* The approach is based on the MJLS approach given in [121]. Another typical approach to compensate for time delays (but no packet losses) is described in [154] where sufficient stability results are obtained for systems with parameter uncertainties. Furthermore, we mention the sequence-based approach [107] that is able to handle nonlinear plants with stochastic data losses and time delays in the network between the actuator and the controller. The model of the sequence-based NCS that is used can be seen as a generalization of the NCS model derived in (own publs. [174,175]). This NCS model is presented in Section 3.3. The virtual control input approach (own publs. [177–179]) also belongs to this category. Here, in contrast to other nominal controller methods, the proposed controller is able to incorporate previously sent and possibly buffered control sequences in the estimation of $\hat{\underline{x}}_{k:k+N-1}$ (see (2.2)). Finally, the stochastic approach in [109] takes the idea of sequence-based control even one step further. Instead of a single control sequence, the controller calculates several sequences, each corresponding to a specific combination of future delay realizations. However, the complexity of the approach exponentially increases with control sequence length and is therefore only suitable for short sequences.

An interesting extension of the basic sequence-based control scheme (2.2) is presented in [152,167]. Here, the nominal controller gain $\mathbf{L}$ is not fixed but varies over the prediction horizon, for example:

$$\underline{U}_k = \begin{bmatrix} \underline{u}_{k|k} \\ \underline{u}_{k|k+1} \\ \vdots \\ \underline{u}_{k+N-1|k} \end{bmatrix} = \begin{bmatrix} \mathbf{L}_0\, \underline{x}_k \\ \mathbf{L}_1\, \hat{\underline{x}}_{k+1} \\ \vdots \\ \mathbf{L}_{k+N-1}\, \hat{\underline{x}}_{k+N-1} \end{bmatrix} . \tag{2.3}$$

We refer to this method as the *generalized nominal controller method.* A practical approach to the method is described in [136] where experimental studies are also carried out. Assuming direct state observations, it has been shown in [167] that sequence-based controllers based on this approach can be transformed to a special kind of TDS. In contrast to standard TDS approaches (see Section 2.3.1), the controller gain also depends on the time delays of the controller-actuator network. Stability criteria for stochastic time delays and losses are presented in [164]. Additionally, if the gain matrices $\mathbf{L}_0, \ldots, \mathbf{L}_{N-1}$ are chosen in a specific way, the resulting controller can also be interpreted as an (unconstrained) sequence-based MPC controller [166].

2.4.3. Methods Based on Model Predictive Control

Another class of sequence-based control methods originates in MPC theory [19]. This is an intuitive connection, as an MPC controller already computes a sequence of optimized control inputs at each time step. In standard MPC, only the first entry of this sequence is applied to the plant and the rest is discarded. A sequence-based MPC controller can therefore very easily be constructed by simply not discarding the rest of the sequence, but adding it to the data packet sent to the actuator.

In the basic MPC method, the control objective is formulated in terms of a cost function. The control inputs are then computed by means of numerical minimization of the cost function over a finite horizon at *each time step.* The procedure is therefore also referred to as receding horizon control. An advantage of this scheme is that state and/or input constraints can be explicitly considered in the minimization. For example, considering a linear quadratic cost function with weighting matrices $\mathbf{Q}$ and $\mathbf{R}$, the MPC controller solves the following possibly constrained (open-loop) optimization problem at each time step k

$$\begin{bmatrix} \underline{u}^*_{k|k} \\ \vdots \\ \underline{u}^*_{k+N-1|k} \end{bmatrix} = \underset{\underline{u}_{k|k},\ \ldots,\ \underline{u}_{k+N-1|k}}{\operatorname{argmin}} \ \mathrm{E}\left\{ \sum_{l=k}^{N-1} \underline{x}_{l|k}^{\top} \mathbf{Q}_l \underline{x}_{l|k} + \underline{u}_{l|k}^{\top} \mathbf{R}_l \underline{u}_{l|k} \right\} , \quad (2.4)$$

where $\underline{x}_{l|k}$ represents the *open-loop* prediction of the system state at time step l based on the information given at time k [2]. While in standard MPC only $\underline{u}^*_{k|k}$ is used and directly applied to the plant, the sequence-based MPC approach utilizes the whole sequence

$$\underline{U}_k = \begin{bmatrix} \underline{u}^*_{k|k} \\ \vdots \\ \underline{u}^*_{k+N-1|k} \end{bmatrix} \tag{2.5}$$

and sends it as one data packet to the actuator.

It is worth clarifying that although the cost function is minimized, the MPC scheme is not a closed-loop optimal solution for the underlying control problem. First, at each optimization step, a shortened horizon is considered instead of the real operation horizon of the control task. Second, in the presence of stochastic disturbances, the real closed-loop optimization problem (see also Section 2.4.4) is replaced by an open-loop approximation. This means that (2.4) is solved considering only the current information about the system state ignoring possible future measurements. Hence, a possible *dual effect* [10] is not taken into account. Finally, in the sequence-based setup, the interactions between the sequences are not considered in the optimization problem. This can be seen in (2.4) as the minimization is performed over the single control inputs $\underline{u}_{k|k}, \ldots, \underline{u}_{k+N-1|k}$ and not over the control sequences $\underline{U}_k, \underline{U}_{k+1}, \ldots$ actually sent to the actuator. As a consequence, the controller-actuator network and the buffering logic are also ignored in the computation of the control sequences.

In summary, the difference between the methods discussed in the previous section and the sequence-based MPC is that the latter does not use a nominal controller to compute control sequences. Instead, the sequences are obtained *as a whole* by online optimization where the buffering scheme and interactions between sequences are neglected.

[2]The double index also indicates that the optimization problem has to be solved at every time step again.

State-of-the-Art

Early work on sequence-based MPC can be found in [68, 71, 72], in which the idea has been elaborated for unconstrained SISO (Single-Input-Single-Output) systems. Currently, there is also a well-developed theory for constrained nonlinear systems with direct state measurements and packet losses [99, 111, 113]. The seminal work [111] derives stabilizing conditions for systems where disturbances and consecutive packet losses are bounded. The approach has been extended to i.i.d. packet losses [110], to stochastic disturbances and quantization effects [108], to correlated packet losses [112], and to piecewise-continuous plants [79]. In this context, the important concept of *prediction consistency* has also been formulated [33, 44]. If prediction consistency holds, stability properties of the standard MPC can be easily transferred to sequence-based MPC.

In addition, MPC schemes with a variable optimization horizon have been proposed in order to adapt the sequence lengths to the network traffic [134]. Stability results for this kind of approaches were first formulated in [135]. This variable horizon approach has also been combined with the idea of not imposing stabilizing terminal constraints [42, 43, 115].

Moreover, the queuing method described in Section 2.3.1 and Section 2.4.2 can also be used in the context of sequence-based MPC. Indeed, the general idea of sequence-based control was originally proposed as an extension of the queuing method to handle control sequences instead of single control inputs [11]. In this respect, [11] resembles the nominal-controller based approach [105] previously described. Also, the works [102, 103] can be interpreted as the MPC version of the nominal controller method [70] that retransmits the queued content in every control packet. In the corresponding MPC schemes [102, 103], however, the authors approach the problem by solving a preconditioned reduced horizon optimal control problem at each time step.

An interesting extension of the basic sequence-based MPC scheme is presented in [95], where the usage of *multiple descriptions* of a certain control sequence is proposed. The multiple descriptions are sent in several partially redundant data packets to the actuator. The more packets the actuator receives, the more future inputs can be retrieved from the multiple descriptions.

2.4.4. Methods Based on Stochastic Optimal Control

Finally, we describe the class of stochastic optimal control approaches to sequence-based control. Similar to sequence-based MPC, these controllers generate control sequences by minimizing a cost function. However, whereas the sequences are optimized independently at each time step in sequenced-based MPC, stochastic optimal control approaches optimize all sequences at the same time. Hence, the interactions between the sequences are also considered and, in particular, possible network-induced delays, packet losses, and the buffering mechanism are explicitly incorporated into the optimization.

To clarify the approach, let us consider a quadratic cost function with weighting matrices $\mathbf{Q}$ and $\mathbf{R}$. The control sequences are then computed by solving the following stochastic closed-loop optimization problem[3]

$$\begin{bmatrix} \underline{U}_0^* \\ \vdots \\ \underline{U}_K^* \end{bmatrix} = \underset{\underline{U}_{0:K}}{\operatorname{argmin}} \operatorname{E}\left\{ \sum_{k=0}^{K} \underline{x}_k^\top \mathbf{Q} \underline{x}_k + \underline{u}_k^\top \mathbf{R} \underline{u}_k \right\} , \tag{2.6}$$

where $\underline{x}_k$ is the system state and $\underline{u}_k$ the control input applied by the actuator. In constrast to the sequence-based MPC optimization problem (2.4), the optimization is carried out *over all control sequences* $\underline{U}_{0:K}$ from the initial time to the terminal time K of the control task. Hence, the interactions between the sequences are fully captured by (2.6). In addition, if (2.6) is optimally solved, the minimizing sequences $\underline{U}_{0:K}^*$ are functions that depend on the information at the corresponding time, i.e.,

$$\underline{U}_k^* = \mu_k^*(\mathcal{I}_k) , \tag{2.7}$$

where $\mathcal{I}_k$ represents the available information at time step k and μ_k^* is the control algorithm used at k. Once the functions $\mu_0^*(\cdot), \ldots, \mu_K^*(\cdot)$ are obtained, they only have to be evaluated during run time in order to calculate the corresponding control sequence.

The sequence-based MPC approach described in Section 2.4.3 can be interpreted as an open-loop approximation of the optimization problem (2.6). Reasons for this have already been discussed in the previous section. Here,

[3]The detailed formulation of the optimization problem is given in Section 3.2.

it is important to note that the sequence-based MPC approaches do not calculate the underlying optimal control law $\mu_0^*(\cdot), \ldots, \mu_K^*(\cdot)$ but directly optimize control inputs of a single control sequence $\underline{u}_{k|k}^*, \ldots, \underline{u}_{k+N-1|k}^*$. Therefore, sequence-based MPC controllers solve the approximated optimization problem (2.4) at each time step numerically, which can be a very time-consuming task. Stochastic optimal controllers solve the harder closed-loop optimization problem (2.6), which has to be done only once.

Having said this, the question arises: Why not always use a stochastic optimal controller instead of a sequence-based MPC-controller? The difficulty is that stochastic optimal control problems such as (2.6) belong to the class of optimization problems that are extremely hard to solve analytically [14]. In particular, if we are confronted with constrained and/or nonlinear systems, there is often no way to solve (2.6) and the use of approximate solutions such as MPC is a good choice. However, for the linear, unconstrained case, it has been shown that for non-sequence-based controllers with quadratic cost function and i.i.d. Gaussian distributed system disturbances, the optimal solution can be derived in *closed-form* [58]. The resulting control law is known as LQG controller, the discovery of which has had a great influence on modern control theory.

In the next chapter, the standard LQG controller is extended to the sequence-based NCS setup with time-varying transmission delays and packet losses. The derived optimal sequence-based controller is referred to as S-LQG and presented in Section 3. In this context, the idea of stochastic closed-loop optimization is clarified and an overview over state-of-the-art optimal control methods for NCS is presented.

3. Optimal Sequence-Based LQG Control

In this chapter, we derive an optimal sequence-based controller that belongs to the class of stochastic optimal control approaches that are described in Section 2.4.4. The results obtained were first published in [172, 174, 175] (own publications). Here, these results are presented in edited form within a unified framework that (hopefully) clarifies many of the mathematical details. The structure of the NCS under consideration is shown in Fig. 3.1. The setup is introduced in detail in the next section but, here, we already want to point out that the NCS comprises a linear plant that is perturbed by Gaussian process and measurement noise. The control objective is to minimize a quadratic cost function that depends on the system state and the applied control inputs. If we neglect the data networks and forget about the sequence-based idea, the system setup exactly corresponds to one of the most fundamental optimal control problems encountered in feedback control. That is the LQG control problem. The problem was solved by Kalman in 1960 [58], whose findings have had a great influence on modern control theory [8]. Here, we revisit the LQG control problem in the context of NCS with time-varying transmission delays and packet losses. In particular, we use the sequence-based method to compensate for the network-induced effects and derive the optimal sequence-based LQG controller. We refer to the resulting controller also as the S-LQG controller.

In the context of NCS, the LQG control problem has also been investigated before. First approaches had already been developed during the 1990s [67, 90]. However, these approaches are not sequence-based and do not consider packet losses, but only time-varying delays. Work [55, 120, 162] that considered the complementary setup with packet losses but no time delays also had a significant impact. Around the same time, the first

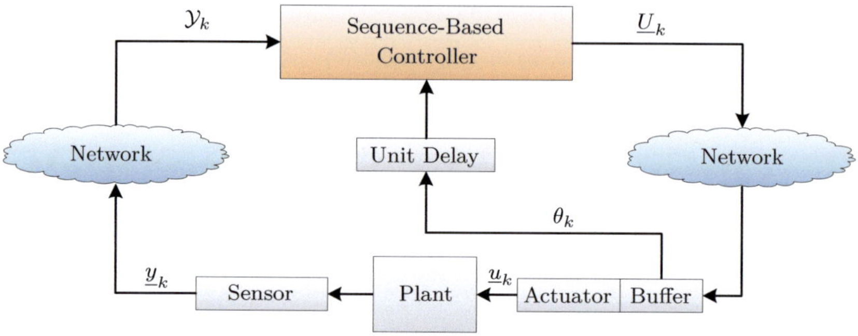

Figure 3.1.: NCS under consideration with sequence-based controller

optimal sequence-based LQG approach to compensate for packet losses was presented [45]. The authors use the concept of a TCP-like network that was also used in [55, 120] before. Yet, in [45] only packet losses are considered and no time delays. Five years later, the approach was extended by [83] in order to incorporate time delays, whereby the resulting controller is suboptimal due to approximations in the course of the control law derivation. Finally, the S-LQG presented closes this gap and constitutes the optimal solution for sequence-based NCS with time-varying transmission delays and packet losses under the assumption of a TCP-like network connection.

It is important to point out that a TCP-like protocol is an information-theoretic concept that is only vaguely related to the real TCP/IP protocol [55]. The similarity of both protocols is given in the use of acknowledgment messages to signal that a data packet has successfully been received. However, in contrast to real TCP/IP networks, where acknowledgment signals can be subject to delays or get lost, it is assumed in a TCP-like protocol that acknowledgment signals are definitely received within the same time step as these are sent. Another difference is that a TCP-like protocol does not initiate a retransmission of data packets that have not been acknowledged by the receiver site. Therefore, the TCP-like protocol can be seen as a means to provide the sender with instantaneous information about the transmission status of data packets.

In practice, a TCP-like network behavior can be realized within a prioritization-based network by assigning acknowledgment signals a higher priority than other data packets, for example. In addition, acknowledgments could be sent redundantly with a higher transmission rate. In wireless networks, different power levels could also be used. Nevertheless, analysis of TCP-like networks is also interesting from the theoretical point of view in its own right. In particular, it allows insight into the far more complex case of using delayed or no acknowledgment signals, and constitutes an upper bound on the performance with these network schemes.

In Section 3.6.2, we will extend the results to a so called UDP-like protocol, which is named after the well-known UDP/IP protocol, as it does not provide acknowledgment signals at all. The advantage of UDP-like networks is that these are much easier to implement in practice. However, the maximum achievable performance is lower than with a TCP-like protocol. Also, optimal controller design for UDP-like networks is much more complicated than for the TCP-like case. This is also the reason why the controller presented in Section 3.6.2 is not optimal in the LQG sense but an approximate solution based on the results obtained in the following sections for TCP-like networks.

Before going into the details of sequence-based controller design for TCP-like networks, we want to illustrate the challenges of this problem class and give a hypothesis as to why the solution to this problem has not been obtained before. To that end, let us consider the case of [45], where only packet losses can occur. In Fig. 3.2, a scenario with three subsequently sent control sequences is depicted. Sequence 1 was sent first and Sequence 3 last. Each of the sequences contains four control inputs where the first one is applicable at the transmission time of the corresponding sequence and the following entries at the subsequent time steps. In the representation, inputs applicable at the same time are vertically aligned. If we assume that Sequence 1 and Sequence 3 have arrived without delay and Sequence 2 got lost, then the indicated control inputs are applied according to their number. Note that only in the case of packet losses, a control sequence can either be lost or received immediately. When received, control inputs of that sequence are always applied beginning with the first entry. Also, due to the previously mentioned TCP-like network protocol, the controller is always informed of which control input has been applied in the previous time step. Hence, when the controller calculates a new control sequence,

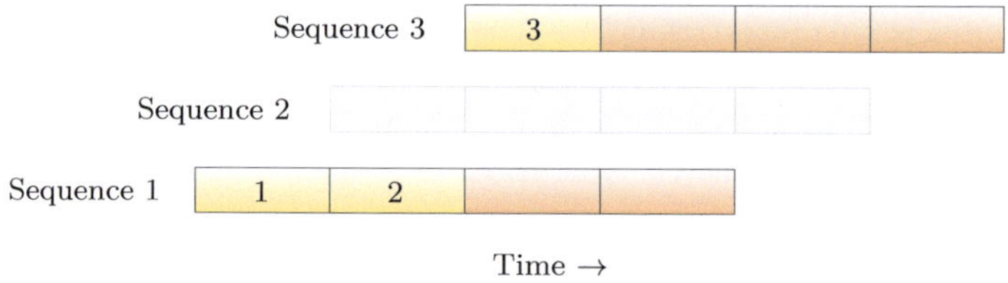

Figure 3.2.: Example for applied control inputs with packet losses only

and we assume that this new sequence will not get lost, then the controller knows exactly which control inputs will have been applied before the calculation is performed for each control input of this sequence. This is important as it means that the controller can calculate the control sequences on a *deterministic* basis. Of course, the assumption that the newly calculated sequence will not get lost might be wrong, but then the calculated control inputs are irrelevant either way as they will not be applied.

Now, we consider the case depicted in Fig. 3.3, in which packet losses and time delays occur simultaneously. In this example, we assume that Sequence 1 arrived without delay, Sequence 2 suffered a delay of one time step, and Sequence 3 got lost. In this scenario, a control sequence might arrive with delay and is then applied starting with the corresponding control input. However, this control input does not have to be the first entry of its sequence. This leads to the problem that despite the TCP-like protocol, the controller only has *stochastic* knowledge of which control inputs will have been applied before the newly calculated inputs might be applied. Hence, the controller can no longer be synthesized on a deterministic basis anymore, and a stochastic controller is needed to take these uncertainties into account.

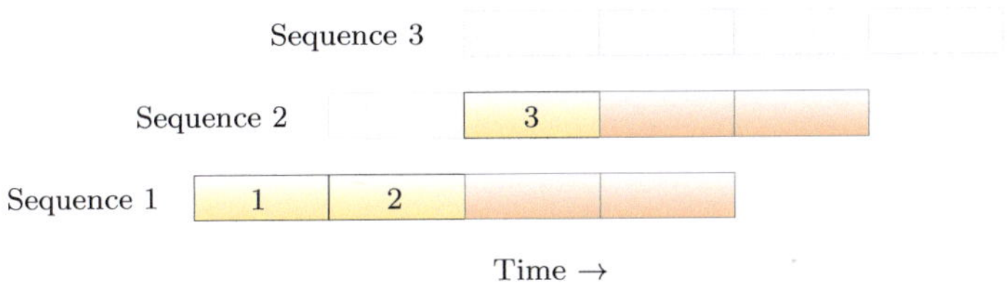

Figure 3.3.: Example of applied control inputs with losses and time delays

In the following sections, a stochastic S-LQG approach is presented that solves the difficulty described above on a stochastic basis. We start with a detailed introduction of the system setup in Section 3.1 and then explicitly define the sequence-based LQG optimization problem in Section 3.2. A sequence-based NCS model is derived in Section 3.3 that not only comprises the buffering scheme but also captures the stochastic nature of the problem illustrated in Fig. 3.3. The solution to the S-LQG is then presented in Section 3.4 and demonstrated in Monte Carlo simulations. Finally, a stability analysis is performed in Section 3.5 and variations of the basic algorithm, such as for UDP-like networks, are presented in Section 3.6.

3.1. NCS Setup under Consideration

The system setup under consideration is shown in Fig. 3.1. The depicted components of the NCS operate on a time-triggered basis, and it is assumed that the clocks of the controller, the sensor, and the actuator are synchronized. The plant is linear and evolves according to

$$\underline{x}_{k+1} = \mathbf{A}_k \underline{x}_k + \mathbf{B}_k \underline{u}_k + \underline{w}_k \ , \qquad \text{for} \quad k \in \mathbb{N}_0 \qquad (3.1)$$

$$\underline{y}_k = \mathbf{C}_k \underline{x}_k + \underline{v}_k \ , \qquad \text{for} \quad k \in \mathbb{N}_{>0} \qquad (3.2)$$

where k is the discrete time index and $\underline{x}_k \in \mathbb{R}^{n_x}$, $\underline{u}_k \in \mathbb{R}^{n_u}$, and $\underline{y}_k \in \mathbb{R}^{n_y}$ denote the plant state, the control input applied by the actuator, and the measured output, respectively. The matrices $\mathbf{A}_k$, $\mathbf{B}_k$, and $\mathbf{C}_k$ are of appropriate dimensions and supposed to be known. The terms $\underline{w}_k \in \mathbb{R}^{n_x}$ and $\underline{v}_k \in \mathbb{R}^{n_y}$ represent mutually independent, zero-mean, Gaussian random processes with finite covariance matrices $\mathbf{W}_k$ and $\mathbf{V}_k$. Likewise, the initial state $\underline{x}_0$ is Gaussian distributed with

$$\overline{\underline{x}}_0 \stackrel{\text{def}}{=} \mathrm{E}\{\underline{x}_0\} \ , \tag{3.3}$$

$$\mathbf{\Lambda}_0 \stackrel{\text{def}}{=} \mathrm{E}\left\{(\underline{x}_0 - \overline{\underline{x}}_0)(\underline{x}_0 - \overline{\underline{x}}_0)^\top\right\} \ . \tag{3.4}$$

The data networks connecting the controller and the actuator (CA-link), as well as the sensor and the controller (SC-link), are subject to stochastically varying transmission delays and stochastic packet losses. The network effects are described by discrete random processes $\tau_k^{CA} \in \mathbb{N}_0$ and $\tau_k^{SC} \in \mathbb{N}_0$ that specify how many time steps a data packet will be delayed if sent at time step k over the CA-link and the SC-link, respectively.

Assumption 3.1 We assume that the processes τ_k^{CA} and τ_k^{SC} are mutually independent, white stationary processes and that their characteristics are known.

The assumption of stationary network characteristics is appropriate for invariable environments and network topologies. Time-variant and/or correlated network characteristics can also be considered within the proposed optimal approach similar to [91] where an extended Gilbert-Elliott model [27, 38] is applied. Here, however, we focus on time-invariant networks for reasons of clarity.

Due to Assumption 3.1, the probability that a packet is delayed by $i \in \mathbb{N}_0$ time steps, for example, does not depend on preceding delay realizations and, hence, is time-invariant. We denote this delay probability by $q^{CA}(i)$ for the controller-actuator network and by $q^{SC}(i)$ for the sensor-controller connection, i.e.,

$$q^{CA}(i) \stackrel{\text{def}}{=} \mathrm{Prob}\left(\tau_k^{CA} = i\right) \ , \tag{3.5}$$

$$q^{SC}(i) \stackrel{\text{def}}{=} \mathrm{Prob}\left(\tau_k^{SC} = i\right) \ . \tag{3.6}$$

Within this description, packet losses correspond to infinite time delays and their probability of occurrence is denoted by $q^{CA}(\infty)$ and $q^{SC}(\infty)$, respectively.

We make the following assumptions in regard to the data networks:

Assumption 3.2 Data packets are marked with time stamps such that τ_k^{CA} and τ_k^{SC} can be recovered at the receiver site.

Assumption 3.3 Sent data fit into one data packet and are not split into several individually routed packets.

Assumption 3.4 The payload of the packets is sufficiently large for effects due to data quantization to be neglected.

Assumption 3.5 The controller-actuator network provides a TCP-like protocol [55, 120], i.e., data packets that are successfully transmitted to the actuator are acknowledged at the controller within the same time step. (See also the introduction to this chapter)

Due to time delays in the sensor-controller network, the controller can receive none, one, or multiple measurements per time step. We denote the set of measurements received at time step $k \in \mathbb{N}_{>0}$ by

$$\mathcal{Y}_k \stackrel{\text{def}}{=} \left\{ \underline{y}_m : m \in \{1, \ldots, k\}, m + \tau_m^{SC} = k \right\} . \tag{3.7}$$

Based on $\mathcal{Y}_k$ and the acknowledgment signal received, the controller computes the control sequence $\underline{U}_k$ of length $N \in \mathbb{N}_{>0}$. The control sequence consists of control inputs $\underline{u}_{k+m|k}$ with $m \in \{0, 1, \ldots, N-1\}$ such that

$$\underline{U}_k \stackrel{\text{def}}{=} \begin{bmatrix} \underline{u}_{k|k}^\top & \underline{u}_{k+1|k}^\top & \cdots & \underline{u}_{k+N-1|k}^\top \end{bmatrix}^\top , \tag{3.8}$$

where an index (k+m|k) expresses that a control input is applicable at time step $k+m$ and was computed at time step k.

In analogy to the controller, the actuator may receive none, one, or multiple data packets at each time step due to the network delays. However, in contrast to the controller, the actuator will only keep the data packet carrying the most recent information among all received packets, i.e., the packet that was generated most recently. This packet is stored in a buffer and all other packets are discarded. Hence, the scheme is also referred to as *past packets rejection logic* [161]. Note that we do not use the rejection logic for data received over the sensor-controller network as so called out-of-sequence measurements still contain useful information. At each time step, the actuator applies the time-consistent control input from the buffered sequence to the plant. It may happen that the actuator does not receive a new admissible data packet before the last control input of the buffered sequence has been applied. In this case, the actuator applies a default control input $\underline{u}_k^{df}$. There are several possibilities for how to choose $\underline{u}_k^{df}$ [119]. Here, we assume that a *zero-input* strategy is used, i.e.,

$$\underline{u}_k^{df} = \underline{0} \ . \tag{3.9}$$

Another widespread choice is the zero-order-hold strategy that will be discussed in Section 3.6.1. Finally, we assume that no control inputs are buffered before operation such that the buffer is initially empty.

The actuator procedure described above can be formalized by

$$\underline{u}_k = \underline{u}_{k|k-\theta_k} \ , \tag{3.10}$$

$$\theta_k \stackrel{\text{def}}{=} \min\Big(\big\{n \in \mathbb{N}_0 : m + \tau_m^{CA} = k - n, \quad m \in \mathbb{N}_0\big\} \cup \{N\}\Big) \ , \tag{3.11}$$

$$\underline{u}_{k|k-N} \stackrel{\text{def}}{=} \underline{u}_k^{df} \ . \tag{3.12}$$

In the next section, we will demonstrate that θ_k can be interpreted as the *age* of the sequence buffered by the actuator. In this context, age means the difference between the time step of sequence generation and the current time step. The value of θ_k is random, as it depends on the random variable τ_k^{CA}. Therefore, the applied control input $\underline{u}_k$ is also a random variable because it depends on θ_k. However, an important implication of Assumption 3.5 is that at time step k, the controller has access to the value of θ_{k-1} and, hence, can reconstruct $\underline{u}_{k-1}$.

Before we introduce the sequence-based LQG control problem in the next section, we will briefly clarify the timing scheme of a control cycle. A control cycle starts with the sensor taking a measurement $\underline{y}_k$ of the system state $\underline{x}_k$ and dispatching it as a time-stamped data packet into the sending queue of the sensor-controller network. This step is skipped at the initial time step $k = 0$. Then, the controller reads the received data queued at the controller site of the SC-link to obtain $\mathcal{Y}_k$ that could already contain $\underline{y}_k$ in case $\tau_k^{SC} = 0$. This is also done at the initial time step where $\mathcal{Y}_0 = \emptyset$. The controller uses $\mathcal{Y}_{0:k}$ to calculate the control sequence $\underline{U}_k$ and writes it into the sending queue of the controller-actuator network. The actuator reads the data received over the CA-link (that already contains $\underline{U}_k$ if $\tau_k^{CA} = 0$) and updates the actuator buffer if necessary. The corresponding control input is applied from the buffer to the plant ending a control cycle.

3.2. Problem Formulation

To formulate the optimal sequence-based LQG control problem, we first characterize the information structure of the given NCS setup in more detail. Particular attention is being paid to the information available to the controller for calculating control sequences. This information consists of all received measurements and all received acknowledgment signals as well as the initial condition and all previous control sequences sent to the actuator. Denoting this information set at time step k by $\mathcal{I}_k$, it holds that

$$\begin{aligned} \mathcal{I}_0 &\stackrel{\text{def}}{=} \{\overline{\underline{x}}_0, \mathbf{\Lambda}_0\} \ , \\ \mathcal{I}_k &\stackrel{\text{def}}{=} \left\{\mathcal{Y}_{1:k}, \underline{U}_{0:k-1}, \theta_{0:k-1}, \overline{\underline{x}}_0, \mathbf{\Lambda}_0\right\} \ , \quad \text{for} \quad k \in \mathbb{N}_{>0} \ . \end{aligned} \tag{3.13}$$

Remark 3.1 As described in the previous section, the information available to the controller also comprises the plant dynamics (3.1), the measurement model (3.2), the initial condition of the actuator buffer, and the stochastic characteristics of disturbances and networks. For simplicity, we do not explicitly state this time-invariant information in the sets $\mathcal{I}_k$ above.

Based on these information sets, we give two definitions that introduce the important concept of admissible control laws.

Definition 3.1 With $\mu_k(\cdot)$ denoting the algorithm used by the controller at time step k, and with $K \in \mathbb{N}_0$ denoting the terminal time of the considered control task, the *control law* is defined as the set of functions $\{\mu_0(\cdot), \ldots, \mu_{K-1}(\cdot)\}$.

Definition 3.2 A control law is called *admissible* if the controller computes a control sequence exclusively based on the information available at that time. This means that a control law is admissible if it holds for all $k \in \{0, \ldots, K-1\}$

$$\underline{U}_k = \mu_k(\mathcal{I}_k) \ . \tag{3.14}$$

In optimization-based control approaches, an admissible control law is derived by minimizing a given cost function that indicates the control performance. The cumulative cost function considered throughout this chapter is an equivalent of the costs considered in the LQG control problem. The cost function is quadratic and given by

$$C_{0\to K}\left(\underline{U}_{0:K-1}\right)$$

$$\stackrel{\text{def}}{=} \underset{\substack{\tau^{CA}_{0:K-1}\\ \tau^{SC}_{1:K}\\ \underline{w}_{0:K-1}\\ \underline{v}_{0:K}}}{\mathrm{E}} \left\{ \underline{x}_K^\top \mathbf{Q}_K \underline{x}_K + \sum_{k=0}^{K-1} \underline{x}_k^\top \mathbf{Q}_k \underline{x}_k + \underline{u}_k^\top \mathbf{R}_k \underline{u}_k \middle| \mathcal{I}_0, \ \underline{U}_{0:K-1} \right\} \tag{3.15}$$

$$= \underset{\substack{\underline{x}_{0:K}\\ \underline{u}_{0:K-1}}}{\mathrm{E}} \left\{ \underline{x}_K^\top \mathbf{Q}_K \underline{x}_K + \sum_{k=0}^{K-1} \underline{x}_k^\top \mathbf{Q}_k \underline{x}_k + \underline{u}_k^\top \mathbf{R}_k \underline{u}_k \middle| \mathcal{I}_0, \ \underline{U}_{0:K-1} \right\} , \tag{3.16}$$

where $K \in \mathbb{N}_{>0}$ denotes the terminal time step when the control task ends. The weighting matrices $\mathbf{Q}_k$ and $\mathbf{R}_k$ are design parameters and positive semidefinite and positive definite, respectively. The expectation is calculated with respect to all occurring random variables as indicated in (3.15). This is equal to taking the expectation only over $\underline{x}_{0:K}$ and $\underline{u}_{0:K-1}$ as these depend on the former. The term $C_{0\to K}$ is referred to as the expected cumulative costs of the control task starting at $k = 0$ and ending at $k = K$. With the notation $C_{0\to K}\left(\underline{U}_{0:K-1}\right)$, we emphasize that the expected cumulative costs can be interpreted as a functional depending on the control sequences $\underline{U}_{0:K-1}$.

Remark 3.2 Throughout this work, we will refer to $C_{0\to K}(\underline{U}_{0:K-1})$ as a cost function instead of cost functional to be consistent with the customary use in the NCS literature.

By minimizing the LQG cost function, the resulting controller tries to hold the system state near the origin of the state space and, hence, pursues to stabilize the plant. The more complex problems that arise when the plant has to be stabilized around a non-zero set point or has to follow a given reference trajectory is discussed in Chapter 4.

Finally, we can formulate the optimization problem to find the optimal admissible control law that minimizes the cost function (3.16), i.e.,

$$C^*_{0\to K} \overset{\text{def}}{=} \min_{\substack{\underline{U}_0=\mu_0(\mathcal{I}_0)\\ \vdots \\ \underline{U}_{K-1}=\mu_{K-1}(\mathcal{I}_{K-1})}} C_{0\to K}\left(\underline{U}_{0:K-1}\right) . \tag{3.17}$$

The term $C^*_{0\to K}$ is referred to as the optimal expected cumulative costs. Concluding this section, we summarize the resulting optimization problem.

Problem 3.1 S-LQG Control Problem (with Zero-Input Strategy)

$$\min_{\underline{U}_{0:K-1}} C_{0\to K}\left(\underline{U}_{0:K-1}\right) ,$$

subject to

$$\begin{aligned}
\underline{x}_{k+1} &= \mathbf{A}_k\underline{x}_k + \mathbf{B}_k\underline{u}_k + \underline{w}_k , \\
\underline{y}_k &= \mathbf{C}_k\underline{x}_k + \underline{v}_k , \\
\underline{u}_k &= \underline{u}_{k|k-\theta_k} , \\
\theta_k &= \min\Big(\left\{ n \in \mathbb{N}_0 : m + \tau^{CA}_m = k - n, \quad m \in \mathbb{N}_0 \right\} \cup \{N\} \Big) , \\
\underline{u}_{k|k-N} &= \underline{u}^{df}_k = \underline{0} , \\
\underline{U}_k &= \mu_k(\mathcal{I}_k) .
\end{aligned}$$

To solve this optimization problem, we need a model of the NCS that adequately captures the characteristics of the sequence-based method and allows for application of the dynamic programming procedure. The derivation of such a model is the subject of the next section.

3.3. Derivation of the Sequence-Based System Model

In this section, we derive the sequence-based NCS model that we use to solve the finite-horizon S-LQG control problem 3.1. First, we derive a model for the controller-actuator network in combination with the actuator. This is achieved by using a state augmentation technique together with a probabilistic formulation of the actuator procedure. Then, the complete NCS model is formulated by the integration of the plant dynamics.

Stochastic Network-Actuator Model

As introduced in (3.10)-(3.12), the control input applied by the actuator at time step k is described by $\underline{u}_k = \underline{u}_{k|k-\theta_k}$ with

$$\begin{aligned} \underline{u}_k &= \underline{u}_{k|k-\theta_k} , \\ \theta_k &= \min\left(\left\{n \in \mathbb{N}_0 : m + \tau_m^{CA} = k - n, m \in \mathbb{N}_0\right\} \cup \{N\}\right) , \\ \underline{u}_{k|k-N} &= \underline{u}_k^{df} . \end{aligned}$$

It can be seen that the random variable θ_k can only take values in the finite set

$$\mathbb{J} \stackrel{\text{def}}{=} \{0, 1, 2, \ldots, N\} .$$

The relationship of θ_k and the actuator output $\underline{u}_k$ is illustrated in Fig. 3.4 which is an example with three control sequences ($\underline{U}_{k-2}$, $\underline{U}_{k-1}$, and $\underline{U}_k$) of length $N = 2$. Moreover, as a result of the past packets rejection logic (governed by (3.11)), it holds for any realizations $i, j, m, n \in \mathbb{J}$ of θ_k

$$\operatorname{Prob}(\theta_{k+1} = j | \theta_0 = m, \theta_1 = n, \ldots, \theta_k = i) = \operatorname{Prob}(\theta_{k+1} = j | \theta_k = i) .$$

Therefore, the evolution of θ_k can be expressed in terms of a first-order Markov-chain with transition matrix $\mathbf{T}$ such that

$$\begin{bmatrix} \operatorname{Prob}(\theta_{k+1} = 0) \\ \operatorname{Prob}(\theta_{k+1} = 1) \\ \vdots \\ \operatorname{Prob}(\theta_{k+1} = N) \end{bmatrix} = \mathbf{T}^\top \begin{bmatrix} \operatorname{Prob}(\theta_k = 0) \\ \operatorname{Prob}(\theta_k = 1) \\ \vdots \\ \operatorname{Prob}(\theta_k = N) \end{bmatrix} . \tag{3.18}$$

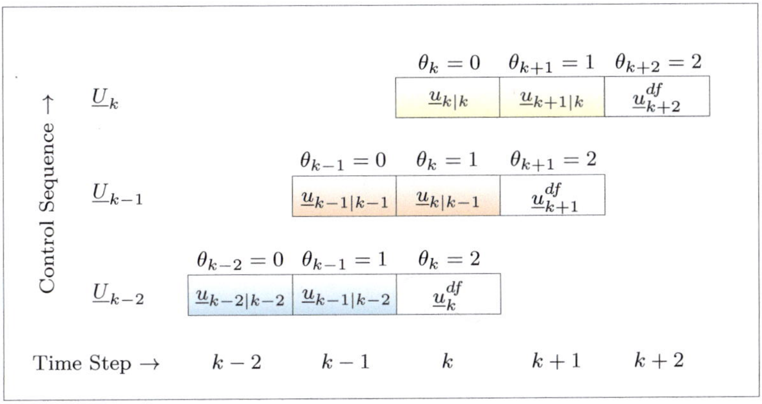

Figure 3.4.: Relation of acknowledgment signal and actuator output in sequence-based NCS model: Shown is the schematic representation of the possible buffer content with three sequences of length $N = 2$ each. A rectangle represents one entry of a data packet. The default control input $\underline{u}^{df}_k$ is added after the end of each data packet. A certain entry is applied by the actuator if θ takes the value specified over the rectangle.

In the following Lemma 3.1, it is shown that the transition matrix is of the form

$$\mathbf{T} = \begin{bmatrix} p(0,0) & p(0,1) & 0 & 0 & \cdots & 0 \\ p(1,0) & p(1,1) & p(1,2) & 0 & \cdots & 0 \\ p(2,0) & p(2,1) & p(2,2) & p(2,3) & \ddots & \vdots \\ \vdots & \vdots & \vdots & \vdots & \ddots & 0 \\ \vdots & \vdots & \vdots & \vdots & & p(N-1,N) \\ p(N,0) & p(N,1) & p(N,2) & p(N,3) & \cdots & p(N,N) \end{bmatrix}, \quad (3.19)$$

where the $p(i,j)$ are the transition probabilities $\text{Prob}\left(\theta_{k+1} = j \,|\, \theta_k = i\right)$. The entries of the transition matrix (and likewise the Markov-chain) are time-invariant due to Assumption 3.1 and can be computed as follows.

Lemma 3.1 The entries of the transition matrix $\mathbf{T}$ describing the evolution of the Markov-chain θ_k (3.11) are given by

$$p(i,j) = \begin{cases} 0 & \text{for } j \geq i+2 \ , \\ 1 - \sum_{n=0}^{i} q^{CA}(n) & \text{for } j = i+1 \ , \\ q^{CA}(j) & \text{for } j < i \leq N \ , \\ 1 - \sum_{n=0}^{N-1} q^{CA}(n) & \text{for } j = i = N \ . \end{cases}$$

The term $q^{CA}(i)$ is defined in (3.5) and describes the probability that a packet sent from the controller to the actuator at time step k will arrive after $\underline{u}_{k+i-1}$ has been applied by the actuator but before $\underline{u}_{k+i}$ is applied.

PROOF. There are four different situations that can occur at the actuator, leading to four different groups of entries $p(i,j)$ of the transition matrix.

- Impossible transitions ($j \geq i+2$):
  The entries of the upper triangle of $\mathbf{T}$ describe transitions from $\theta_k = i$ to $\theta_{k+1} \geq i+2$. Such transitions occur with probability zero as θ_k can only increase by one per time step due to the buffering scheme.

- Buffered sequence is not replaced ($j = i+1$):
  These upper diagonal entries of $\mathbf{T}$ describe the probability that θ_k will increase by one, which corresponds to the case that the buffered sequence is not replaced by another sequence at time step k. This only occurs if the actuator does not receive a packet that was generated after the currently buffered sequence. It therefore holds

  $$p(i,i+1) = \prod_{m=0}^{i} \left(1 - \check{q}^{CA}(m)\right) \ , \tag{3.20}$$

  where $\check{q}^{CA}(m)$ is the probability that a packet generated m time steps ago will be received during the next time step. This probability can be calculated by

  $$\check{q}^{CA}(m) = q^{CA}(m) \cdot \left(1 - \sum_{n=0}^{m-1} q^{CA}(n)\right)^{-1} \ . \tag{3.21}$$

The intuition behind (3.21) is that the a-priori probability that a sequence will suffer a delay of m time steps is modified by the knowledge that the sequence has not been received yet ($\tau_{k-m}^{CA} \geq m$). Therefore, the second term in (3.21) normalizes $q^{CA}(m)$ over the remaining delay probability mass of that sequence. Using (3.21) in (3.20) results in

$$\begin{aligned} p(i, i+1) &= \prod_{m=0}^{i} \left(1 - \check{q}^{CA}(m)\right) \\ &= \prod_{m=0}^{i} \left(1 - q^{CA}(m) \cdot \left(1 - \sum_{n=0}^{m-1} q^{CA}(n)\right)^{-1}\right) \\ &= \prod_{m=0}^{i} \left(\frac{1 - \sum_{n=0}^{m} q^{CA}(n)}{1 - \sum_{n=0}^{m-1} q^{CA}(n)}\right) \\ &= 1 - \sum_{n=0}^{i} q^{CA}(n) \ . \end{aligned} \tag{3.22}$$

- Buffered sequence is replaced ($j \leq i < N$):
  The lower triangle of $\mathbf{T}$ describes transitions where θ_k does not increase. Hence, the buffered sequence is replaced by a newer one. The probability of such an event is the probability that a new packet is received (that was generated after the currently buffered sequence) while none of the packets are received that were generated after this new packet, i.e.,

$$p(i, j) = \check{q}^{CA}(j) \cdot \prod_{m=0}^{j-1} \left(1 - \check{q}^{CA}(m)\right) \ . \tag{3.23}$$

The probabilities involved in (3.23) have been conditioned on the fact that all corresponding packets have not yet been received. This condition ensures that the transition is valid, i.e., that it can start in $\theta_k = i$. Using (3.21) and (3.22) in (3.23) results in

$$\begin{aligned} p(i,j) &= \check{q}^{CA}(j) \cdot \prod_{m=0}^{j-1} \left(1 - \check{q}^{CA}(m)\right) \\ &= q^{CA}(j) \cdot \left(1 - \sum_{n=0}^{j-1} q^{CA}(n)\right)^{-1} \cdot \left(1 - \sum_{m=0}^{j-1} q^{CA}(m)\right) \\ &= q^{CA}(j) \ . \end{aligned} \tag{3.24}$$

- Empty Buffer ($j = i = N$):
  The entry describes the probability that the actuator buffer was empty during the last time step and will remain empty for at least one more time step. This corresponds to the probability that no admissible packet will be received in the next time step conditioned on the fact that no relevant packet has yet been received. Hence,

$$p(N,N) = \prod_{m=0}^{N-1} \left(1 - \check{q}^{CA}(m)\right) \overset{(3.22)}{=} 1 - \sum_{n=0}^{N-1} q^{CA}(n) \ . \tag{3.25}$$

□

To link θ_k with the output of the actuator, we introduce the vector

$$\underline{\rho}_k = \begin{bmatrix} \left[\underline{u}_{k|k-1}^{\top} \;\; \underline{u}_{k+1|k-1}^{\top} \;\; \cdots \;\; \underline{u}_{k+N-3|k-1}^{\top} \;\; \underline{u}_{k+N-2|k-1}^{\top}\right]^{\top} \\ \left[\underline{u}_{k|k-2}^{\top} \;\; \underline{u}_{k+1|k-2}^{\top} \;\; \cdots \;\; \underline{u}_{k+N-3|k-2}^{\top}\right]^{\top} \\ \vdots \\ \left[\underline{u}_{k|k-N+2}^{\top} \;\; \underline{u}_{k+1|k-N+2}^{\top}\right]^{\top} \\ \underline{u}_{k|k-N+1} \end{bmatrix}^{\top} , \tag{3.26}$$

with $\underline{\rho}_k \in \mathbb{R}^{n_\rho}$ and $n_\rho = n_u \cdot \sum_{i=1}^{N-1} i = n_u \cdot N \cdot (N-1)/2$. This vector contains all control inputs of previously sent sequences that could still be applied to the plant $\left(\underline{U}_{k-1}, \ldots, \underline{U}_{k-N+1}\right)$. This is illustrated in Fig. 3.5, where the relevant control sequences are depicted for the case of $N = 3$.

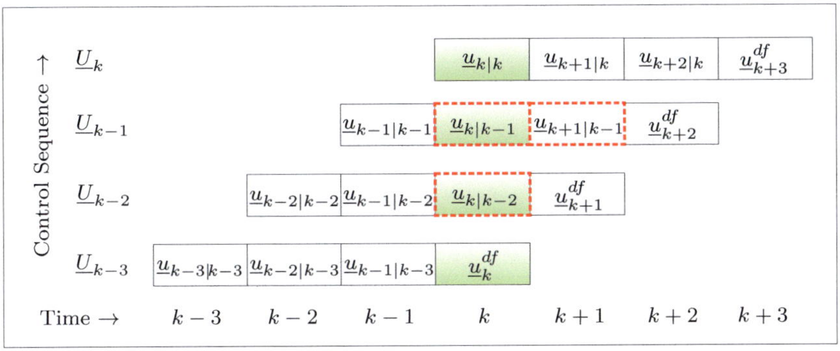

Figure 3.5.: Representation of control sequences $\underline{U}_k, \ldots, \underline{U}_{k-3}$, where control inputs corresponding to the same time step are vertically aligned. The default control input $\underline{u}^{df}_k$, which is not part of the sequences, is added to the end of each sequence. All control inputs possibly applied by the actuator at time step k are colored green. Control inputs that are part of $\underline{\rho}_k$ are marked by a red dashed rectangle.

Finally, combining (3.10), (3.12), and Lemma 3.1, leads to the following state space model of the network-actuator system

$$\underline{\rho}_{k+1} = \mathbf{F}\underline{\rho}_k + \mathbf{G}\underline{U}_k \ , \tag{3.27}$$

$$\underline{u}_k = \mathbf{H}_k(\theta_k)\underline{\rho}_k + \mathbf{J}_k(\theta_k)\underline{U}_k + \mathbf{D}_k(\theta_k)\underline{u}^{df}_k \ , \tag{3.28}$$

with

$$\mathbf{F} = \begin{bmatrix} \overbrace{\mathbf{0}}^{n_u} & \overbrace{\mathbf{0}}^{n_u(N-2)} & \overbrace{\mathbf{0}}^{n_u} & \overbrace{\mathbf{0}}^{n_u(N-3)} & \cdots & \overbrace{\mathbf{0}}^{n_u} & \overbrace{\mathbf{0}}^{n_u} \\ \mathbf{0} & \mathbf{I} & \mathbf{0} & \mathbf{0} & \cdots & \mathbf{0} & \mathbf{0} \\ \mathbf{0} & \mathbf{0} & \mathbf{0} & \mathbf{I} & \cdots & \mathbf{0} & \mathbf{0} \\ \vdots & \vdots & \vdots & \vdots & \ddots & \vdots & \vdots \\ \mathbf{0} & \mathbf{0} & \mathbf{0} & \mathbf{0} & \cdots & \mathbf{I} & \mathbf{0} \end{bmatrix} \begin{matrix} \#rows: \\ \}n_u(N-1) \\ \}n_u(N-2) \\ \}n_u(N-3) \\ \\ \}n_u \end{matrix} ,$$

(#columns: n_u, $n_u(N-2)$, n_u, $n_u(N-3)$, …, n_u, n_u)

$$\delta_{(\theta_k,i)} = \begin{cases} 1\,, & \text{if} \quad \theta_k = i \\ 0\,, & \text{if} \quad \theta_k \neq i \end{cases}\,, \qquad \mathbf{G} = \begin{bmatrix} \overbrace{\mathbf{0}}^{n_u} & \overbrace{\mathbf{I}}^{n_u(N-1)} \\ \mathbf{0} & \mathbf{0} \end{bmatrix} \begin{matrix} \}n_u(N-1) \\ \}\frac{n_u(N-1)(N-2)}{2} \end{matrix}\,,$$

(#columns: n_u, $n_u(N-1)$; #rows: $n_u(N-1)$, $\frac{n_u(N-1)(N-2)}{2}$)

$$\mathbf{J}_k(\theta_k) = \Big[\, \overbrace{\delta_{(\theta_k,0)}\mathbf{I}}^{n_u} \quad \overbrace{\mathbf{0}}^{n_u(N-1)} \,\Big]\,, \qquad \mathbf{D}_k(\theta_k) = \overbrace{\delta_{(\theta_k,N)}\mathbf{I}}^{n_u}\,,$$

(#columns: n_u, $n_u(N-1)$)

$$\mathbf{H}_k(\theta_k) = \Big[\, \overbrace{\delta_{(\theta_k,1)}\mathbf{I}}^{n_u} \quad \overbrace{\mathbf{0}}^{n_u(N-2)} \quad \overbrace{\delta_{(\theta_k,2)}\mathbf{I}}^{n_u} \quad \overbrace{\mathbf{0}}^{n_u(N-3)} \quad \cdots \quad \overbrace{\delta_{(\theta_k,N-1)}\mathbf{I}}^{n_u} \,\Big]\,,$$

(#columns: n_u, $n_u(N-2)$, n_u, $n_u(N-3)$, …, n_u)

Dynamic Model of the Complete NCS

To combine the stochastic network-actuator model with the plant, we introduce the augmented state vector

$$\underline{\xi}_k = \begin{bmatrix} \underline{x}_k \\ \underline{\rho}_k \end{bmatrix}\,,$$

with $\underline{\xi}_k \in \mathbb{R}^{n_\xi}$ and $n_\xi = n_x + n_\rho$. Using the augmented state, we can combine (3.1), (3.2), (3.27), and (3.28) to get the following model of the open loop system

$$\underline{\xi}_{k+1} = \begin{bmatrix} \mathbf{A}_k & \mathbf{B}_k \cdot \mathbf{H}_k(\theta_k) \\ \mathbf{0} & \mathbf{F} \end{bmatrix} \underline{\xi}_k + \begin{bmatrix} \mathbf{B}_k \cdot \mathbf{J}_k(\theta_k) \\ \mathbf{G} \end{bmatrix} \underline{U}_k + \mathbf{D}_k(\theta_k)\underline{u}_k^{df} + \begin{pmatrix} \underline{w}_k \\ \underline{0} \end{pmatrix}$$

and

$$\underline{y}_k = \begin{bmatrix} \mathbf{C}_k & \mathbf{0} \end{bmatrix} \underline{\xi}_k + \underline{v}_k\,.$$

Setting $\underline{u}_k^{df} = \underline{0}$ (zero-input strategy), we summarize the derived NCS model.

Sequence-Based NCS Model (with Zero-Input Strategy)

$$\begin{aligned} \underline{\xi}_{k+1} &= \widehat{\mathbf{A}}_k(\theta_k)\underline{\xi}_k + \widehat{\mathbf{B}}_k(\theta_k)\underline{U}_k + \mathbf{D}_k(\theta_k)\underline{u}_k^{df} + \underline{\widehat{w}}_k \ , \\ \underline{y}_k &= \widehat{\mathbf{C}}_k\underline{\xi}_k + \underline{v}_k \ , \end{aligned} \tag{3.29}$$

with

$$\widehat{\mathbf{A}}_k(\theta_k) = \begin{bmatrix} \mathbf{A}_k & \mathbf{B}_k \cdot \mathbf{H}_k(\theta_k) \\ \mathbf{0} & \mathbf{F} \end{bmatrix} , \qquad \widehat{\mathbf{C}}_k = \begin{bmatrix} \mathbf{C}_k & \mathbf{0} \end{bmatrix} ,$$

$$\widehat{\mathbf{B}}_k(\theta_k) = \begin{bmatrix} \mathbf{B}_k \cdot \mathbf{J}_k(\theta_k) \\ \mathbf{G} \end{bmatrix} , \quad \underline{\widehat{w}}_k = \begin{pmatrix} \underline{w}_k \\ \underline{0} \end{pmatrix} , \quad \underline{\xi}_k = \begin{bmatrix} \underline{x}_k \\ \underline{\rho}_k \end{bmatrix} .$$

The open-loop system model described by (3.29) is a Markov Jump Linear System (MJLS) as it consists of a set of linear models that are stochastically switched according to a Markov chain. A general overview over this system class can be found in [141], for example. In this context, θ_k is also referred to as the *mode* of the MJLS. The MJLS (3.29) is non-homogenous due to the stochastic disturbances and the control inputs. It is important to point out that in contrast to [141], here, the mode is only available with a delay of one time step and measurements are subject to time-varying transmission delays and packet losses. Also, we do not assume a special structure of the filter and the controller. The results obtained in the former work can therefore not be applied to solve Problem 3.1.

Only recently, a similar model for sequence-based NCS was published in [107]. The authors start with a more general setup and then derive an equivalent model of (3.29) as a special case. Therefore, the general NCS model of [107] can be seen as an extension of (3.29).

3.4. The Optimal Solution

Using the system model (3.29) derived in the last section, we solve the finite-horizon sequence-based optimization problem 3.1. This is performed in detail in Appendix A via the dynamic programming procedure for stochastic optimal control problems [13]. This control law obtained is referred to as Sequence-Based Linear Quadratic Gaussian (S-LQG) control in analogy to the standard LQG control. In the derivation of the control law, we assume that a zero-input strategy is used, i.e., $\underline{u}_k^{df} = \underline{0}$. Another choice for the default control input is discussed in Section 3.6.1. In the following, we first summarize the results derived in Appendix A on the optimal S-LQG controller and discuss important aspects of the control law. Then, the S-LQG is illustrated by means of a Monte Carlo simulation.

3.4.1. Main Result

Theorem 3.1 *Consider the optimization problem 3.1. Then,*

a) as in standard LQG control, the separation principle holds, i.e., the optimal control law can be separated into

1) an estimator that calculates the minimum mean squared error (MMSE) estimate $\mathrm{E}\{\underline{\xi}_k|\mathcal{I}_k\}$ *of the augmented state, and into*

2) an optimal state feedback controller with gain matrix $\mathbf{L}_k(\theta_{k-1})$*,*

b) the feedback matrix $\mathbf{L}_k(\theta_{k-1})$ *explicitly depends on* θ_{k-1} *and, hence, on the sequence buffered by the actuator at time step* $k-1$ *,*

c) the optimal control law is linear in the MMSE estimate of the augmented state such that

$$\underline{U}_k = \mathbf{L}_k(\theta_{k-1}) \cdot \mathrm{E}\left\{\underline{\xi}_k \middle| \mathcal{I}_k\right\} , \tag{3.30}$$

d) for all $i \in \mathbb{J} = \{0, \ldots, N\}$ the feedback matrix $\mathbf{L}_k(\theta_{k-1})$ is given by

$$\begin{aligned}\mathbf{L}_k(i) = &- \left(\sum_{j=0}^{N} p(i,j) \left[\widehat{\mathbf{R}}_k(j) + \widehat{\mathbf{B}}_k(j)^\top \mathbf{K}_{k+1}(j) \widehat{\mathbf{B}}_k(j) \right] \right)^\dagger \\ &\cdot \left(\sum_{j=0}^{N} p(i,j) \left[\widehat{\mathbf{B}}_k(j)^\top \mathbf{K}_{k+1}(j) \widehat{\mathbf{A}}_k(j) \right] \right) ,\end{aligned} \tag{3.31}$$

where $\widehat{\mathbf{A}}_k(j)$, $\widehat{\mathbf{B}}_k(j)$, and $\widehat{\mathbf{R}}_k(j)$ are defined in (3.29) and the matrix $\mathbf{K}_{k+1}(j)$ can be computed by the recursion

$$\begin{aligned}\mathbf{K}_k(i) = &\sum_{j=0}^{N} p(i,j) \left[\widehat{\mathbf{Q}}_k(j) + \widehat{\mathbf{A}}_k(j)^\top \mathbf{K}_{k+1}(j) \widehat{\mathbf{A}}_k(j) \right] \\ &- \left(\sum_{j=0}^{N} p(i,j) \left[\widehat{\mathbf{A}}_k(j)^\top \mathbf{K}_{k+1}(j) \widehat{\mathbf{B}}_k(j) \right] \right) \\ &\cdot \left(\sum_{j=0}^{N} p(i,j) \left[\widehat{\mathbf{R}}_k(j) + \widehat{\mathbf{B}}_k(j)^\top \mathbf{K}_{k+1}(j) \widehat{\mathbf{B}}_k(j) \right] \right)^\dagger \\ &\cdot \left(\sum_{j=0}^{N} p(i,j) \left[\widehat{\mathbf{B}}_k(j)^\top \mathbf{K}_{k+1}(j) \widehat{\mathbf{A}}_k(j) \right] \right) ,\end{aligned} \tag{3.32}$$

that is initialized with $\mathbf{K}_K(i) = \begin{bmatrix} \mathbf{Q}_K & \mathbf{0} \\ \mathbf{0} & \mathbf{0} \end{bmatrix}$ for all $i \in \mathbb{J}$.

PROOF. The proof is given in Appendix A. □

Theorem 3.1 implies that the optimal sequence-based controller can be computed by first solving the recursion (3.32) in order to compute the feedback gain (3.31). Then, the optimal control sequence is calculated according to (3.30) by multiplication of the feedback matrix with the MMSE estimate $\mathrm{E}\{\underline{\xi}_k | \mathcal{I}_k\}$. The calculation of the MMSE estimate has already been intensively studied in the literature [82, 118]. We will review these results after discussing the general structure of the controller.

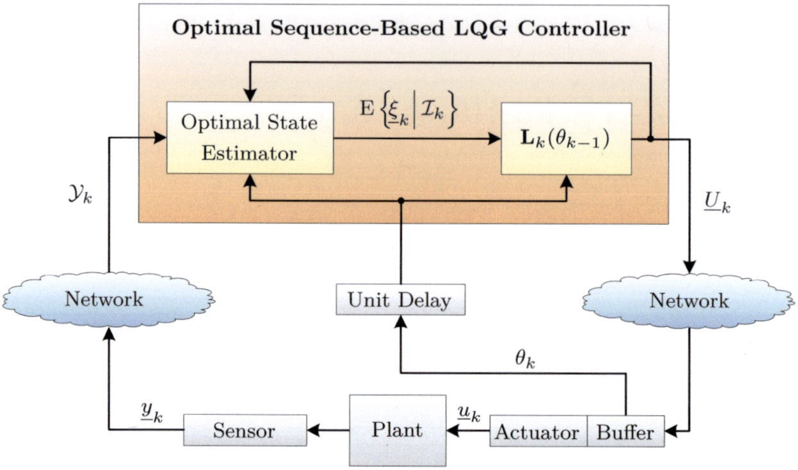

Figure 3.6.: Structure of the optimal sequence-based LQG controller

The overall structure of the S-LQG is depicted in Fig. 3.6. Due to the validity of the separation principle, the controller can be decomposed into the optimal MMSE estimator and the optimal sequence-based feedback controller. The latter is identical to the optimal sequence-based controller that would be obtained if the state of the plant were directly accessible. In this respect, it can be interpreted as the sequence-based equivalent to the well-known LQR (Linear Quadratic Regulator). It is important to point out that the depicted idealized acknowledgment signals of the TCP-like connection (see Assumption 3.5) are crucial for the separation property to hold. Similar observations have been made in [55, 120] for the non-sequence-based setup with packet losses. Theorem 3.1 extends these results to the sequence-based case.

An advantage of the S-LQG is that the gain matrices $\mathbf{L}_k(j)$ can be computed offline for all time steps $k = 0, \ldots, K-1$ and for all values of $j \in \mathbb{J}$. During run time, only the corresponding matrix must be chosen according to the current time step and the value of the last acknowledgment signal. This allows the use of the S-LQG in applications with fast time constants such as motion control. Furthermore, (3.31) reveals a strong relationship to standard LQG control and, in particular, the recursive equation $\mathbf{K}_k(i)$ is reminiscent of the standard Riccati equation [58]. In Section 3.5, it is shown that, similar to the standard Riccati equation, the corresponding recursion (3.32) converges to a steady state for long time horizons under appropriate assumptions and depends on the network characteristics. When convergence occurs, this leads to a huge reduction of the required memory space needed to store the gain matrices. Especially, in practical applications where memory might be rare, this is an extremely important result.

Moreover, not only can the control law be derived analytically but the expected cumulative costs $C^*_{0\to K}$ induced by the S-LQG can as well. This result is formalized in the next theorem.

Theorem 3.2 *The minimal expected cumulative costs* $C^*_{0\to K}$ *defined in* (3.17) *are given by*

$$\begin{aligned} C^*_{0\to K} = \mathrm{E}\left\{\underline{\xi}_0^\top \mathbf{K}_0(N)\underline{\xi}_0 \middle| \mathcal{I}_0\right\} + \sum_{k=0}^{K-1} \mathrm{E}\left\{\underline{e}_k^\top \mathbf{P}_k \underline{e}_k \middle| \mathcal{I}_0\right\} \\ + \sum_{k=0}^{K-1} \mathrm{E}\left\{\underline{\widehat{w}}_k^\top \mathbf{K}_{k+1}(\theta_k)\underline{\widehat{w}}_k \middle| \mathcal{I}_0\right\} , \end{aligned} \tag{3.33}$$

with

$$\begin{aligned} \underline{e}_k &\overset{\text{def}}{=} \underline{\xi}_k - \mathrm{E}\left\{\underline{\xi}_k \middle| \mathcal{I}_k\right\} , \\ \mathbf{P}_k &\overset{\text{def}}{=} \mathrm{E}\left\{\widehat{\mathbf{Q}}_k(\theta_k) + \widehat{\mathbf{A}}_k(\theta_k)^\top \mathbf{K}_{k+1}(\theta_k)\widehat{\mathbf{A}}_k(\theta_k) \middle| \mathcal{I}_k\right\} - \mathbf{K}_k(\theta_{k-1}) , \\ \mathbf{K}_k(\theta_{k-1}) &= \mathrm{E}\left\{\widehat{\mathbf{Q}}_k(\theta_k) + \widehat{\mathbf{A}}_k(\theta_k)^\top \mathbf{K}_{k+1}(\theta_k)\widehat{\mathbf{A}}_k(\theta_k) \middle| \mathcal{I}_k\right\} \\ &\quad - \mathrm{E}\left\{\widehat{\mathbf{A}}_k(\theta_k)^\top \mathbf{K}_{k+1}(\theta_k)\widehat{\mathbf{B}}_k(\theta_k) \middle| \mathcal{I}_k\right\} \\ &\quad \cdot \mathrm{E}\left\{\widehat{\mathbf{R}}_k(\theta_k) + \widehat{\mathbf{B}}_k(\theta_k)^\top \mathbf{K}_{k+1}(\theta_k)\widehat{\mathbf{B}}_k(\theta_k) \middle| \mathcal{I}_k\right\}^\dagger \\ &\quad \cdot \mathrm{E}\left\{\widehat{\mathbf{B}}_k(\theta_k)^\top \mathbf{K}_{k+1}(\theta_k)\widehat{\mathbf{A}}_k(\theta_k) \middle| \mathcal{I}_k\right\} . \end{aligned}$$

PROOF. The proof is given in Appendix A. □

Theorem 3.2 quantifies that the expected costs of the S-LQG can theoretically be calculated offline. In practice, however, it is easier to use bounds on the costs as the terms $\mathrm{E}\left\{\underline{e}_k^\top \mathbf{P}_k \underline{e}_k \middle| \mathcal{I}_k\right\}$ are extremely difficult to evaluate [126]. The costs are further analyzed in Section 3.5 with respect to the stability of the S-LQG. Moreover, in Chapter 5, an event-triggered controller is presented that evaluates the costs online for different scenarios.

Finally, we will briefly review how the MMSE estimate of the augmented state $\mathrm{E}\{\underline{\xi}_k|\mathcal{I}_k\}$ can be calculated. It has been shown in [118] that $\mathrm{E}\{\underline{\xi}_k|\mathcal{I}_k\}$ can be obtained by a time-varying Kalman filter that is extended by a buffer to account for time delays in the sensor-controller network. The buffer is used to store the measurements received. In case a delayed measurement is received out-of-sequence, it is sorted into the buffer at the correct position. Then the state estimate is obtained by recalculating the measurement history. The chosen length of the buffer, denoted by N^B, has to be sufficiently large to ensure that the true MMSE is obtained. In particular, for optimal results it is sufficient if

$$N^B = \max\left\{i : \ i \in \mathbb{N}_{>0}, \ q^{SC}(i) > 0\right\}. \tag{3.34}$$

This ensures that all measurements can be buffered that are needed to incorporate every possible out-of-sequence measurement into the state estimation. However, measurements that have suffered a long time delay usually have only a small influence on the state estimate. In practice, it is therefore reasonable to save computing resources by limiting the buffer length and treating N^B as a design parameter.

Another design parameter is the length N of the control sequences. In the next section, we investigate the influence of this parameter on the control performance in a simulation with a double integrator plant. In Section 3.5, we will also derive general guidelines for the choice of N by analyzing the stability properties of the closed-loop system.

3.4.2. Evaluation

We perform simulations of the S-LQG with a classical double integrator plant to give a sense of the controller and to compare it to state-of-the-art approaches. The simulated setup is shown in Fig. 3.7. The

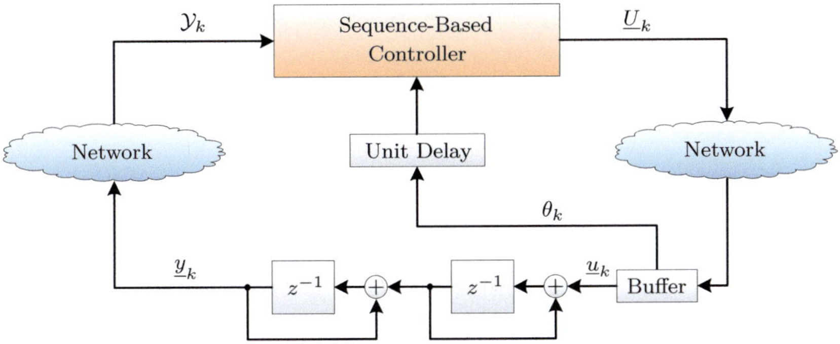

Figure 3.7.: Sequence-based NCS setup for simulation with double integrator plant

double integrator frequently occurs in practical control problems [114]. For example, it can be interpreted as an accelerated mass where the position and the velocity of the mass are the states of the system and the accelerating force is the control input. The model of the double integrator is given by

$$\mathbf{A} = \begin{bmatrix} 1 & 1 \\ 0 & 1 \end{bmatrix},$$

$$\mathbf{B} = \begin{bmatrix} 0 \\ 1 \end{bmatrix},$$

$$\mathbf{C} = \begin{bmatrix} 1 & 0 \end{bmatrix}.$$

We compare the proposed S-LQG controller with two other sequence-based LQG approaches. The first one is the optimal approach described in [45] that does not account for delay effects (LQG-Loss). For the second controller, we consider the approach of [83] that incorporates delays, however, based on an approximation (LQG-Approx). In addition, we also implement a conventional LQG controller that is collocated at the plant. In this way, the latter gives a natural lower bound for the minimum achievable costs of the other approaches. To implement the controllers and simulate

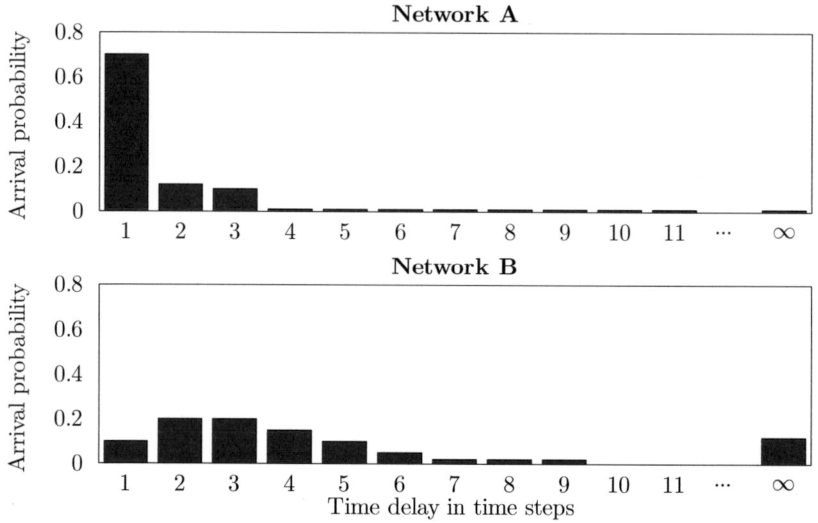

Figure 3.8.: Stochastic network characteristics considered in simulations with double integrator plant

the system, we set the weighting matrices of the cost function (3.16), the noise covariances, and the initial condition to

$$\begin{aligned}
\mathbf{Q}_k &= \mathbf{I} \,, \\
\mathbf{R}_k &= 1 \,, \\
\underline{\overline{x}}_0 &= \begin{bmatrix} 100 \\ 0 \end{bmatrix}, \\
\mathbf{\Lambda}_0 &= \begin{bmatrix} 0.5^2 & 0 \\ 0 & 0.5^2 \end{bmatrix} \,, \\
\mathrm{E}\left\{\underline{w}_k{}^\top \underline{w}_k\right\} &= \begin{bmatrix} 0.1^2 & 0 \\ 0 & 0.1^2 \end{bmatrix}, \\
\mathrm{E}\left\{\underline{v}_k^\top \underline{v}_k\right\} &= 0.2^2 \,.
\end{aligned}$$

We conduct two different simulations with two different network models. The probability density functions of the delay distributions of both networks are shown in Fig. 3.8. We refer to Network A as a *good* network because small time delays occur with high probability. Correspondingly, Network B is considered as a *bad* network because longer delays and packet losses have a higher probability. For controller design, the minimum mean square error estimate $\mathrm{E}\{\underline{\xi}_k|\mathcal{I}_k\}$ is obtained via the filter given in [82]. The length of the used measurement buffer is set to $N^B = 11$ such that the filter yields the optimal estimate. The default control input, which is applied if the actuator buffer runs empty, is set to $\underline{u}_k^{df} = 0$. For each controller and network, and for different control sequence lengths N, 200 Monte Carlo simulation runs are evaluated with 200 time steps each. The average of the attained cumulative costs is determined for each case according to (3.16). The results are plotted in Fig. 3.9.

With the good network, the costs induced by the S-LQG, LQG-Loss, and LQG-Approx do not vary substantially. However, when considering the bad network connections, the difference between the controllers becomes clear. The S-LQG performs significantly better than the LQG-Loss and the LQG-Approx. In particular, for sequence lengths $N > 2$, the costs of the S-LQG are approximately half the costs of the LQG-Loss. The LQG-Approx is not even able to stabilize the system.

It is interesting to note that the induced costs of the optimal control approaches, i.e., S-LQG and LQG-Loss, decrease with increasing sequence lengths. This justifies the optimal sequence-based approach as a general tool to compensate for time delays and packet losses in NCS. The costs also no longer significantly decrease for $N \geq 3$. This can be used as guideline for choosing the sequence length. Here, $N = 3$ would be a good choice. A systematic approach to choosing the sequence length is described in Section 3.5.

3.5. Stability Analysis of the S-LQG Controller

The S-LQG is the optimal controller for the NCS setup under consideration, as no other controller exists that leads to lower expected costs. However, is still unclear whether these costs are reasonably bounded, i.e., whether the S-LQG can stabilize the system. And if it does, which sequence length is required to guarantee stability? To answer these questions, we analyze the costs induced by the S-LQG. One problem is that these costs (3.16)

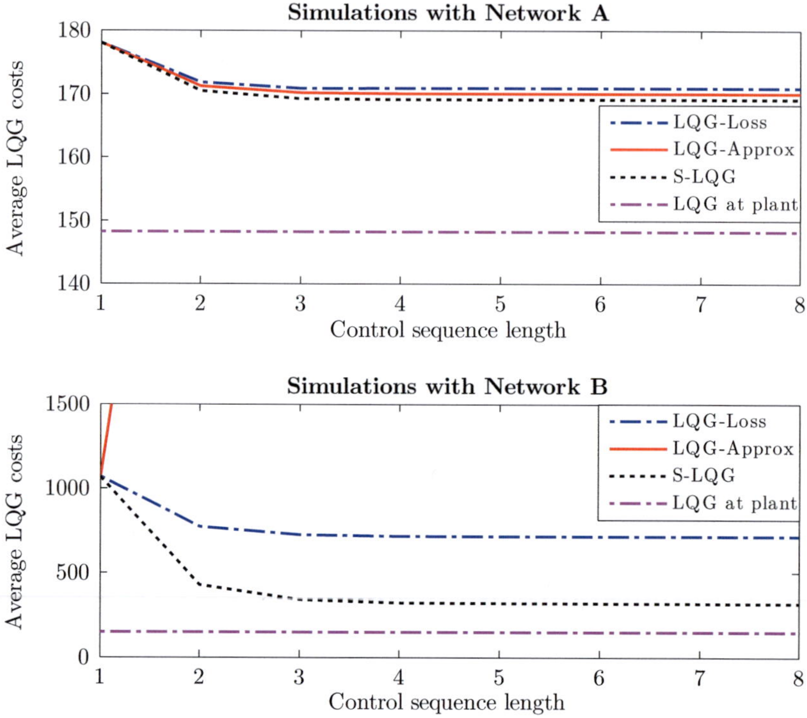

Figure 3.9.: Averaged cumulative costs for both networks and for different control sequence lengths

will grow unbounded with increasing terminal time K due to the presence of stochastic disturbances and unbounded network effects. Therefore, it is more meaningful to analyze the long run average costs instead.

Definition 3.3 The long run average costs C_∞ are defined as

$$C_\infty(\underline{U}_0, \underline{U}_1, \dots) \stackrel{\text{def}}{=} \limsup_{K\to\infty} \frac{1}{K} \cdot C_{0\to K}\left(\underline{U}_{0:K-1}\right) , \tag{3.35}$$

where the expected cumulative costs $C_{0\to K}\left(\underline{U}_{0:K-1}\right)$ are given by (3.16).

The long run average costs represent the costs occurring per time step when the system runs for a long time. In LQG control, the boundedness of the long run average costs is a standard criterion for assessing the stability of the closed-loop system [15]. In particular, if the long run average costs are bounded from above, the system is considered to be stable. We now give a definition of this stability criterion.

Definition 3.4 The system is said to be Long Run Average Costs stable (*LRAC-stable*), if the induced long run average costs $C_\infty(\underline{U}_0, \underline{U}_1, \ldots)$ are bounded from above, i.e., if there exists $\overline{C} \in \mathbb{R}$ such that for all initial conditions

$$C_\infty(\underline{U}_0, \underline{U}_1, \ldots) \leq \overline{C} \ .$$

Otherwise, if no such $\overline{C}$ exists, the system is called Long Run Average Costs unstable (*LRAC-unstable*).

In the context of MJLS, other stochastic stability criteria have also been introduced such as the *almost sure stability*, *stochastic stability* (SS), or *mean square stability* (MSS) [141]. Based on these and other concepts, stability conditions have been derived for constrained systems with directly accessible state [111, 115], for undisturbed systems [33, 41, 70, 105, 135], for systems with bounded disturbances [103], and for NCS where only losses and no time delays occur [79]. Yet, these results are not applicable here because of the substantial differences in the systems in question. In the next section, conditions for the LRAC-(in)stability of the S-LQG are presented based on [172] (own publication). Closest to these results is the work [45, 120], in which conditions have been derived for the non-sequence-based setup and for the case with packet losses only, respectively. The relationship between these conditions is pointed out at the end of Section 3.5.1.

3.5.1. Main Result

To analyze the stability properties of the S-LQG controller, we introduce the following additional assumptions[1].

Assumption 3.6 It holds for the system (3.1), (3.2) and cost function (3.16)

a) the plant is time-invariant, i.e., $\mathbf{A}_k = \mathbf{A}$, $\mathbf{B}_k = \mathbf{B}$, and $\mathbf{C}_k = \mathbf{C}$,

b) $(\mathbf{A}, \mathbf{C})$ is observable,

c) $(\mathbf{A}, \mathbf{Q}^{1/2})$ is observable,

d) $(\mathbf{A}, \mathbf{W}^{1/2})$ is controllable,

e) $\mathbf{V} > 0$.

The main results on the stability analysis are presented in the form of the following two theorems. The first theorem gives a sufficient condition for LRAC-stability and the second a sufficient condition for LRAC-instability. The proofs to these theorems have been moved to Appendix B for better readability.

Theorem 3.3 *Consider the NCS setup described in Section 3.1 with Assumption 3.6 that is controlled by the S-LQG given by Theorem 3.1 . Then, the system is* ***LRAC-stable*** *if both of the following conditions are satisfied:*

a) It holds either that $\max|\text{eig}(\mathbf{A})| < 1$ *or* $q^{SC}_{arr} > q^{SC}_{crit}$ *, where* $q^{SC}_{arr} = \sum_{m=0}^{N^B} q^{SC}(m)$ *is the measurement arrival probability and* q^{SC}_{crit} *is the critical probability that is obtained by the solution of the quasi-convex optimization problem*

$$q^{SC}_{crit} \overset{\text{def}}{=} \underset{q}{\text{argmin}}\ \Psi(\mathbf{Y}, \mathbf{Z}, q) > \mathbf{0}$$

[1]The assumptions could also be relaxed towards detectability, reachability, and positive semidefiniteness. However, this proliferates the mathematical description and, therefore, has been refrained.

with constraints

$$0 \leq \mathbf{Y} \leq \mathbf{I}\,, \quad \Psi\left(\mathbf{Y}, \mathbf{Z}, q\right) = \begin{bmatrix} \mathbf{Y} & \sqrt{q}\left(\mathbf{YA} + \mathbf{ZC}\right) & \sqrt{1-q}\mathbf{YA} \\ (*)^\top & \mathbf{Y} & \mathbf{0} \\ (*)^\top & \mathbf{0} & \mathbf{Y} \end{bmatrix} .$$

b) There exist $N+1$ matrices $\widehat{\mathbf{L}}(0:N)$ and $N+1$ positive definite matrices $\mathbf{X}(0:N)$ such that

$$\mathbf{X}(j) > \sum_{i=0}^{N} p(j,i) \left(\widehat{\mathbf{A}}(i) + \widehat{\mathbf{B}}(i)\widehat{\mathbf{L}}(j)\right)^\top \mathbf{X}(i) \left(\widehat{\mathbf{A}}(i) + \widehat{\mathbf{B}}(i)\widehat{\mathbf{L}}(j)\right), \tag{3.36}$$

PROOF. The proof is given in Appendix B.1. □

Condition a) in Theorem 3.3 is derived in [120] and ensures that the estimation error covariance $\mathrm{E}\left\{\underline{e}_k\underline{e}_k^\top \,\middle|\, \mathcal{I}_0\right\}$ is bounded. This is always the case if all eigenvalues of the system are Lyapunov stable, i.e., are within the unit circle. However, if the plant has eigenvalues outside of the unit circle, then there exists a critical arrival probability q_{crit}^{SC} such that the estimation error can grow unbounded when the probability that a measurement will eventually arrive at the controller q_{arr}^{SC} is smaller than the critical probability. The arrival probability q_{arr}^{SC} can be influenced by the length of the measurement buffer N^B used at the controller to store the measurement history (see Section 3.4).

The second condition in Theorem 3.3 ensures that the control related costs are bounded. The term control related costs refers to the costs that would be induced if the controller had direct access to the plant state, i.e., if the estimation error $\underline{e}_k$ were $\underline{0}$. The expression in Theorem 3.3 b) is derived by bounding the true costs of the system that depend on the time-varying S-LQG controller gain by the costs a time-invariant controller would induce. It is worth pointing out that the condition (3.36) depends on the transition probabilities p and therefore on the sequence length N.

Next, we present a sufficient condition for LRAC-instability. Note that if the S-LQG is not able to stabilize the system, then there will be no other controller that is able to stabilize the system in the LRAC-sense. Therefore, the results presented in the next theorem constitute fundamental bounds on the stabilizability of linear sequence-based NCS.

Theorem 3.4 *Consider the NCS setup described in Section 3.1 with Assumption 3.6 controlled by the S-LQG given in Theorem 3.1 . Then, the system is* ***LRAC-unstable*** *if at least one of the following conditions is satisfied*

a) $q_{arr}^{SC} \leq 1 - (\max|\text{eig}(\mathbf{A})|)^{-2}$,

where $q_{arr}^{SC} = \sum_{m=0}^{N^B} q^{SC}(m)$,

b) $p(N,N) \cdot \max|\text{eig}(\mathbf{A})|^2 > 1$,

where $p(N,N) = 1 - \sum_{m=0}^{N-1} q^{CA}(m)$ *is defined by Lemma 3.1.*

PROOF. The proof is given in Appendix B.2. □

Again, the first condition refers to the estimation error covariance and states that the covariance grows unbounded if the measurement arrival probability of measurements is smaller than an expression that depends on the maximum eigenvalue of the system matrix [118]. The second condition of Theorem 3.4 gives a sufficient condition for the unboundedness of the control related costs. It relates the maximum eigenvalue of the system with the probability that the actuator buffer would run out of applicable control inputs. Hence, the condition depends on the sequence length N.

To get an intuition for the derived criteria, let us consider that no time delays and packet losses occur in the network connections. This is basically the standard LQG setup and condition (3.36) producing

$$\mathbf{X}(0) > \left(\widehat{\mathbf{A}}(0) + \widehat{\mathbf{B}}(0)\widehat{\mathbf{L}}(0)\right)^{\top} \mathbf{X}(0) \left(\widehat{\mathbf{A}}(0) + \widehat{\mathbf{B}}(0)\widehat{\mathbf{L}}(0)\right) . \tag{3.37}$$

According to Lyapunov theory [127], if there exists an $\mathbf{X}(0)$ such that the inequality holds, then all eigenvalues of $(\widehat{\mathbf{A}}(0) + \widehat{\mathbf{B}}(0)\widehat{\mathbf{L}}(0))$ are strictly smaller than one. This implies that the system is Lyapunov stable and one can easily show that this also implies LRAC-stability [15]. Hence, Theorem 3.3 b) is justified in this example. Furthermore, (3.37) always has a solution if $(\widehat{\mathbf{A}}(0), \widehat{\mathbf{B}}(0))$ is stabilizable, because in this case, there is an $\widehat{\mathbf{L}}(0)$ with $\max|\text{eig}(\widehat{\mathbf{A}}(0) + \widehat{\mathbf{B}}(0)\widehat{\mathbf{L}}(0))| < 1$. Observing that the stabilizability of $(\widehat{\mathbf{A}}(0), \widehat{\mathbf{B}}(0))$ is equivalent to the stabilizability of $(\mathbf{A}, \mathbf{B})$ implies that Theorem 3.3 b) reduces to the stabilizability of $(\widehat{\mathbf{A}}(0), \widehat{\mathbf{B}}(0))$ which is known to be a necessary stability condition in standard LQG control.

Another interesting case is when there are only packet losses and no time delays in the network connections. This situation is investigated in [120] for the non-sequence-based setup, i.e., $N = 1$. In this case, the authors show that the corresponding condition in Theorem 3.3 b) is not only sufficient but also necessary. The same setup has also been investigated for the sequence-based case, i.e., $N \geq 1$ [45]. The derived conditions are similar to the results in Theorem 3.3 and Theorem 3.4 . However, we do not need the assumption regarding the steady-state distribution of the Markov chain as needed in Prop. 3 of [45].

In the following, we use Theorem 3.3 and Theorem 3.4 to derive bounds on the critical sequence length N^{crit} required to stabilize a system. Information about N^{crit} is very interesting for practical implementations as it gives an important guideline for choosing the length of the control sequence. However, before we state the results, we need the following lemma that allows for evaluating of Theorem 3.3 b) in terms of a Linear Matrix Inequality (LMI) [17] feasibility problem.

Lemma 3.2 The condition in Theorem 3.3 b) is equivalent to the existence of $N+1$ matrices $\mathbf{Y}(0:N)$ and $N+1$ matrices $\mathbf{Z}(0:N)$ such that

$$\Xi(\mathbf{Y}(0:N), \mathbf{Z}(0:N)) > \mathbf{0} \quad \text{and} \quad \mathbf{0} < \mathbf{Y}(0:N) < \mathbf{I}\,,$$

with

$$\Xi(\mathbf{Y}(0:N), \mathbf{Z}(0:N)) \stackrel{\text{def}}{=} \begin{bmatrix} \mathbf{\Theta}(0) & \mathbf{0} & \cdots & \mathbf{0} \\ \mathbf{0} & \mathbf{\Theta}(1) & & \mathbf{0} \\ \vdots & & \ddots & \vdots \\ \mathbf{0} & \mathbf{0} & \cdots & \mathbf{\Theta}(N) \end{bmatrix},$$

$$\mathbf{\Theta}(j) \stackrel{\text{def}}{=} \begin{bmatrix} \mathbf{Y}(j) & \Sigma(j,0) & \cdots & \Sigma(j,N) \\ \Sigma(j,0)^\top & \mathbf{Y}(0) & & \mathbf{0} \\ \vdots & & \ddots & \vdots \\ \Sigma(j,N)^\top & \mathbf{0} & \cdots & \mathbf{Y}(N) \end{bmatrix},$$

$$\Sigma(j,i) \stackrel{\text{def}}{=} \sqrt{p(j,i)}\left(\mathbf{Y}(j)\widehat{\mathbf{A}}(i)^\top + \mathbf{Z}(j)\widehat{\mathbf{B}}(i)^\top\right) .$$

Proof. The equivalence of the expression can be shown by first applying the Schur complement [17] to (3.36) and then introducing the new variables $\mathbf{Y}(j) = (\mathbf{X}(j))^{-1}$ and $\mathbf{Z}(j) = (\mathbf{X}(j))^{-1} \cdot \widehat{\mathbf{L}}(j)$. The result is the LMI given above. □

Using Lemma 3.2, we can formulate a corollary for computing the critical sequence length that guarantees LRAC-stability.

Corollary 3.1 Denoting the shortest sequence length that guarantees LRAC-stability by N^{crit}, it holds that

$$N^{min} \geq N^{crit} \geq N^{max} , \tag{3.38}$$

where

$$N^{min} = \min_n \left\{ n \in \mathbb{N}_0 : \sum_{m=0}^{n} q^{CA}(m) \geq 1 - \frac{1}{\max|\text{eig}(\mathbf{A})|^2} \right\}, \tag{3.39}$$

with $\max|\text{eig}(\mathbf{A})| \neq 0$. The upper bound N^{max} on the critical sequence length N^{crit} can be obtained as the solution of the optimization problem

$$N^{max} = \underset{N}{\text{argmin}}\ \mathbf{\Xi}(\mathbf{Y}(0:N), \mathbf{Z}(0:N)) > \mathbf{0} , \tag{3.40}$$

with constraints $\mathbf{0} < \mathbf{Y}(0:N) < \mathbf{I}$.

Proof. The corollary is a direct implication of Lemma 3.2, Theorem 3.3 b) and Theorem 3.4 b). □

Finally, we give a result concerning the convergence properties of the S-LQG controller gain matrices.

Theorem 3.5 *If the system is LRAC-stable according to Theorem 3.3 , the recursion* (3.32) *converges for all* $i, j \in \mathbb{J}$ *to the* $N+1$ *matrices* $\overline{\mathbf{K}}(0:N)$ *given by*

$$\begin{aligned}\overline{\mathbf{K}}(i) =& \sum_{j=0}^{N} p(i,j) \left[\widehat{\mathbf{Q}}(j) + \widehat{\mathbf{A}}(j)^\top \overline{\mathbf{K}}(j)\widehat{\mathbf{A}}(j)\right] \\ & - \left(\sum_{j=0}^{N} p(i,j) \left[\widehat{\mathbf{A}}(j)^\top \overline{\mathbf{K}}(j)\widehat{\mathbf{B}}(j)\right]\right) \\ & \cdot \left(\sum_{j=0}^{N} p(i,j) \left[\widehat{\mathbf{R}}(j) + \widehat{\mathbf{B}}(j)^\top \overline{\mathbf{K}}(j)\widehat{\mathbf{B}}(j)\right]\right)^\dagger \\ & \cdot \left(\sum_{j=0}^{N} p(i,j) \left[\widehat{\mathbf{B}}(j)^\top \overline{\mathbf{K}}(j)\widehat{\mathbf{A}}(j)\right]\right) .\end{aligned} \tag{3.41}$$

Proof. The proof is given in Appendix B.3. □

Theorem 3.5 has a very important consequence for the practical application of the S-LQG. The convergence of the controller gain drastically reduces the memory required as only $N+1$ controller gains have to be stored instead of $(N+1) \cdot K$, which would be impossible for the infinite-horizon scenario. The theorem also states that the convergence of the controller gain will occur in all relevant scenarios, i.e., in all scenarios where control makes sense due to the LRAC-stability of the system.

In the next section, we demonstrate the applicability of the derived stability criteria with a numerical example.

3.5.2. Evaluation

In the following simulation, we focus on demonstrating the conditions formulated in Corollary 3.1, Theorem 3.3 b), and Theorem 3.4 b), which refer to the LRAC-stability of the control related costs. The conditions in Theorem 3.3 a) and Theorem 3.4 a) concerning the error covariance of the estimated state are investigated in detail [118]. Therefore, we consider a directly observable plant in which the network between the sensor and the controller is replaced by a direct point-to-point connection. The system (3.1) and (3.2) is chosen to

$$\mathbf{A} = \begin{bmatrix} 0.5 & 0 \\ 1 & 1.5 \end{bmatrix} ,$$
$$\mathbf{B} = \begin{bmatrix} 1 \\ 0 \end{bmatrix} ,$$
$$\mathbf{C} = \mathbf{I} .$$

The system matrix $\mathbf{A}$ has the eigenvalues 0.5 and 1.5. In the simulation, we set the covariances and initial conditions to

$$\mathbf{W} = \mathbf{I} ,$$
$$\mathbf{V} = \mathbf{0} ,$$
$$\underline{\overline{x}}_0 = \begin{bmatrix} 10 & 10 \end{bmatrix} ,$$
$$\mathbf{\Lambda}_0 = \mathbf{I} .$$

We assume that delays occur in the controller-actuator network with uniform distribution according to $q^{CA}(0) = q^{CA}(1) = q^{CA}(2) = q^{CA}(3) = 0.25$, and that there are no packet losses. The controller is computed as described in Theorem 3.1 for the control sequence lengths $N = \{1, 2, 3, 4\}$. The weighting matrices of the cost function are set to $\mathbf{Q} = \mathbf{I}$ and $\mathbf{R} = 10 \cdot \mathbf{I}$.

Evaluating (3.40) of Corollary 3.1 with an LMI-solver such as SeDuMi [131], the optimization problem is infeasible for N={1, 2} and feasible for $N = \{3, 4\}$. The upper bound for the critical sequence length N^{crit} that guarantees LRAC-stability is therefore $N^{max} = 3$. In addition, considering (3.39) of Corollary 3.1, it can be seen that $1 - 1/\max|\mathrm{eig}(\mathbf{A})|^2 \approx 0.55$ and $\sum_{r=0}^{1} q_r^{CA} = 0.5$ and $\sum_{r=0}^{2} q_r^{CA} = 0.75$. This shows that the system is LRAC-unstable for $N \leq 2$ and we have $N^{min} = 3$. Combining these facts, it holds that $3 \leq N^{crit} \leq 3$ and, hence, $N^{crit} = 3$.

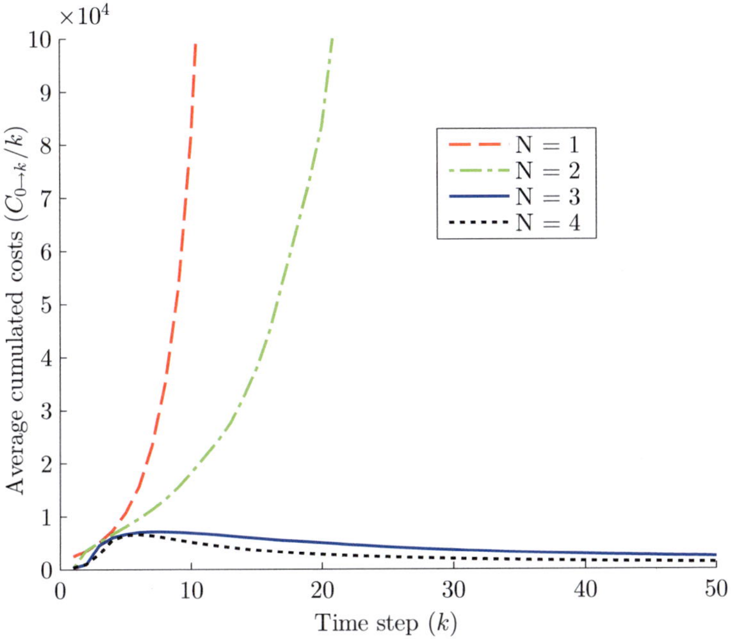

Figure 3.10.: Comparison of the averaged cumulative costs $C_{0\to k}/k$ induced until time step k for different lengths of the control sequences.

To verify this result, we perform Monte Carlo simulations with 10^5 simulation runs over 50 time steps for each of the control sequences $N = \{1, \ldots, 4\}$. The induced costs are divided by the current time step $C_{0\to k}/k$ and averaged over all simulation runs. The result is plotted against the time step as shown in Fig. 3.10. The exponential increasing averaged costs for sequence lengths $N = \{1, 2\}$ indicate that the system is LRAC-unstable for these cases. However, for longer sequences $N = \{3, 4\}$ the costs converge and, hence, are bounded. Therefore, the simulation verifies the theoretical result of $N^{crit} = 3$.

3.6. Variations of the Basic S-LQG Algorithm

In this section, we discuss two variations of the basic S-LQG algorithm. This is first the application of the so called hold-input strategy for choosing the default control input. Second, we investigate the case of UDP-like networks which do not provide any acknowledgment signals.

3.6.1. Hold-Input Strategy

Recall that the default control input $\underline{u}_k^{df}$ is the control input that is applied by the actuator if the actuator buffer runs empty. In the derivation of the S-LQG in Section 3.4, we set the default control input $\underline{u}_k^{df} = \underline{0}$. This strategy is also referred to as *zero-input* strategy. It is a good choice in many scenarios, but might be insufficient in others. In this section, we therefore discuss a very widespread alternative: the *zero-order hold* strategy. In this scheme, the previously applied control input is applied again until new control data are received, i.e., $\underline{u}_k^{df} = \underline{u}_{k-1}$. Note that neither of the strategies is superior to the other in general. It depends on the scenario which scheme performs better [119].

In the zero-input strategy, the default control input is always known by the controller in advance. In the zero-order hold strategy, the default control input changes every time a new sequence is received by the actuator. Due to transmission delays and packet losses in the network, it is not known a-priori which sequence will arrive at the actuator and therefore which will be the last successfully received control sequence by a certain time step. As a consequence, the default control input depends on the time delay realizations τ^{CA} of the controller-actuator network and is not known a-priori. To cope with this effect, we integrate $\underline{u}_k^{df}$ in the augmented system state, i.e.,

$$\underline{\tilde{\xi}}_k \overset{\text{def}}{=} \begin{bmatrix} \underline{x}_k \\ \underline{\rho}_k \\ \underline{u}_k^{df} \end{bmatrix} = \begin{bmatrix} \underline{\xi}_k \\ \underline{u}_k^{df} \end{bmatrix} , \tag{3.42}$$

where $\underline{\widetilde{\xi}}_k \in \mathbb{R}^{n_x+n_\rho+n_u}$ is the new system state and $\underline{x}_k$ and $\underline{\rho}_k$ are given by (3.1) and (3.26), respectively. Thus, the NCS system model derived in Section 3.3 must be adapted according to

Sequence-Based NCS Model (with Zero-Order Hold Strategy)

$$\underline{\widetilde{\xi}}_{k+1} = \widetilde{\mathbf{A}}_k(\theta_k)\,\underline{\widetilde{\xi}}_k + \widetilde{\mathbf{B}}_k(\theta_k)\,\underline{U}_k + \underline{\widetilde{w}}_k \ ,$$
$$\underline{y}_k = \widetilde{\mathbf{C}}_k\,\underline{\widetilde{\xi}}_k + \underline{v}_k \ ,$$

with

$$\widetilde{\mathbf{A}}_k(\theta_k) \stackrel{\text{def}}{=} \begin{bmatrix} \mathbf{A}_k & \mathbf{B}_k \cdot \mathbf{H}_k(\theta_k) & \delta_{(\theta_k,N)} \cdot \mathbf{I} \\ \mathbf{0} & \mathbf{F} & \mathbf{0} \\ \mathbf{0} & \mathbf{H}_k(\theta_k) & \delta_{(\theta_k,N)} \cdot \mathbf{I} \end{bmatrix} , \quad \widetilde{\mathbf{C}}_k \stackrel{\text{def}}{=} \begin{bmatrix} \mathbf{C}_k & \mathbf{0} \end{bmatrix} ,$$

$$\widetilde{\mathbf{B}}_k(\theta_k) \stackrel{\text{def}}{=} \begin{bmatrix} \mathbf{B}_k \cdot \mathbf{J}_k(\theta_k) \\ \mathbf{G} \\ \mathbf{0} \end{bmatrix} , \quad \underline{\widetilde{w}}_k \stackrel{\text{def}}{=} \begin{pmatrix} \underline{w}_k \\ \underline{0} \\ \underline{0} \end{pmatrix} , \quad \underline{\widetilde{\xi}}_k \stackrel{\text{def}}{=} \begin{bmatrix} \underline{x}_k \\ \underline{\rho}_k \\ \underline{u}_k^{df} \end{bmatrix} .$$

Using this model, we can formulate the solution to the optimal sequence-based LQG control problem with a zero-order hold strategy.

Corollary 3.2 Consider the optimization problem 3.1 with the zero-input strategy replaced by the zero-order hold strategy, i.e., $\underline{u}_k^{df} = \underline{u}_{k-1}$, then the optimal solution is given by Theorem 3.1 with $\underline{\xi}_k$, $\widehat{\mathbf{A}}_k(\theta_k)$, $\widehat{\mathbf{B}}_k(\theta_k)$, $\widehat{\mathbf{C}}_k$, and $\underline{\widehat{w}}_k$ are replaced by $\underline{\widetilde{\xi}}_k$, $\widetilde{\mathbf{A}}_k(\theta_k)$, $\widetilde{\mathbf{B}}_k(\theta_k)$, $\widetilde{\mathbf{C}}_k$, and $\underline{\widetilde{w}}_k$, respectively.

Proof. Replacing the corresponding quantities, the proof follows the lines of the proof of Theorem 3.1 exactly. □

3.6.2. Control without Network Acknowledgments

In this section, we briefly discuss the problem of controller design for networks that provide a UDP-like protocol instead of the TCP-like protocol considered thus far. Using a UDP-like protocol, the network does not provide acknowledgments for successfully sent data packets. For the controller-actuator network connection, this implies that Assumption 3.5 does not hold, as the controller has no direct information about the control inputs that are applied by the actuator. In particular, neither the sequence buffered at the previous time step (as indicated by θ_{k-1}), nor the applied control input $\underline{u}_{k-1}$ is perfectly known at time step k. However, for an optimal controller design, these values are needed and, hence, must be estimated based on the information available to the controller.

Unfortunately, the estimation error of these values depends on the control sequences sent. Therefore, a control sequence not only influences the system directly when applied by the actuator, but also indirectly by affecting the future estimation error of the buffered sequence. A lower estimation error allows for better control decisions and, thus, has an indirect effect on the induced costs. This direct and indirect influence of the control is also referred to as the *dual effect* [10]. The dual effect implies that the separation property (see Theorem 3.1 for TCP-like networks) does not hold for UDP-like networks. This means that the optimal control problem cannot be divided into a control problem and into an estimation problem without loss of optimality. In particular, Lemma A.3 does not hold for UDP-like networks as the error covariance $\mathrm{E}\left\{\underline{e}_{K-1}^\top \mathbf{P}_{K-1} \underline{e}_{K-1} \middle| \mathcal{I}_{K-2}\right\}$ depends on $\underline{U}_{K-2}$.

It is widely recognized that optimal control problems involving the dual effect are extremely hard to solve and optimal analytic solutions are not available in general. Therefore, we propose an approximate solution by means of a so called *certainty equivalence* control approach [10]. In this strategy, the controller is first designed assuming that all stochastic system variables that are subject to the dual effect are treated as if these quantities are perfectly known. Then, the corresponding variables are replaced by their MMSE estimates.

Applying this idea to the S-LQG, we exchange the control law

$$\underline{U}_k = \mathbf{L}_k(\theta_{k-1}) \cdot \mathrm{E}\{\underline{\xi}_k | \mathcal{I}_k\}$$

given in (3.30) with the certainty equivalence control law

$$\underline{U}_k = \mathbf{L}_k(\mathrm{E}\{\theta_{k-1} | \mathcal{I}_k\}) \cdot \mathrm{E}\{\underline{\xi}_k | \mathcal{I}_k\} \ . \tag{3.43}$$

This control law is not optimal. However, it provides a feasible and easy structure, and typically performs astonishing well in practice. We refer to this controller as S-LQG-UDP.

Before we can use the S-LQG-UDP, we need an estimator that calculates the estimates $\mathrm{E}\{\theta_{k-1} | \mathcal{I}_k\}$ and $\mathrm{E}\{\underline{\xi}_k | \mathcal{I}_k\}$ for each time step. In NCS literature, however, the problem of state estimation is typically investigated with just one network between the sensor and the controller [118, 126, 129]. For cases in which there is only a network between the controller and the actuator, methods based on the unknown input observer have been proposed [28, 62, 96]. Yet, these filters only exist under rather strong rank conditions on the system matrices, i.e., the system is required to be minimum-phase. In [45, 81, 120], networks between the sensor and the controller, as well as between the controller and the actuator, are taken into account. However, [45] does not account for packet delays and [120] relies on a TCP-like protocol. The approach in [81] could be used in our scenario to estimate θ_k. However, the filter neglects the correlation of the state and the buffered control sequence.

Therefore, in [175] (own publication), an estimator is proposed based on the *Interacting Multiple Model* (IMM) method. The approach accounts for the correlations involved and performs a joint input-state estimation to obtain $\mathrm{E}\{\theta_{k-1} | \mathcal{I}_k\}$ and $\mathrm{E}\{\underline{\xi}_k | \mathcal{I}_k\}$ simultaneously. The key ingredient to estimator is the NCS model derived in Section 3.3. In [175], it is shown that there is a strong relationship between state estimation in predictive NCS (in particular sequence-based NCS) and multi-target tracking. Also, simulations are provided that demonstrate the improved performance of the estimator.

Coming back to the control problem, we apply the estimator described in [175] (own publication) to implement the S-LQG-UDP. In the following, we want to get an idea of the degree of suboptimality that the approach comprises. To that end, we repeat the Monte Carlo simulation described in Section 3.4.2 with the double integrator plant and compare the performance of the S-LQG with the S-LQG-UDP. In the simulation, the S-LQG uses a TCP-like network while the S-LQG-UDP uses a UDP-like network with the same stochastic characteristics. Other simulation parameters are not changed.

The simulation results are depicted in Fig. 3.11. One can see that the S-LQG-UDP does not perform better than the S-LQG. This is expected as the latter uses the additional information provided by the TCP-like network optimally. For Network A, the S-LQG-UDP stabilizes the system with around 35% higher costs than the S-LQG. This leaves room for improvement but also seems acceptable considering that the S-LQG-UDP requires much less communication. For Network B, we can see that the S-LQG-UDP becomes unstable for sequence lengths $N = 1$ and $N = 2$. The effective packet loss rate when using these length is more than 90% and 70%, respectively (see Fig. 3.8). As the S-LQG stabilizes the plant with a TCP-like network, this demonstrates the value of the acknowledgment signals, in particular, in the presence of massive packet losses and long time delays. For sequence lengths $N \geq 3$, the S-LQG-UDP stabilizes the system with approximately 150% higher costs than the S-LQG. Unfortunately, we do not know the value of the minimal achievable costs that an optimal controller for a UDP-like network would induce. Hence, we cannot determine how well the S-LQG-UDP utilizes the available information. However, the plots clearly indicate that the S-LQG-UDP provides better control quality with longer sequence lengths. This is an important observation as the controller is based on an approximation.

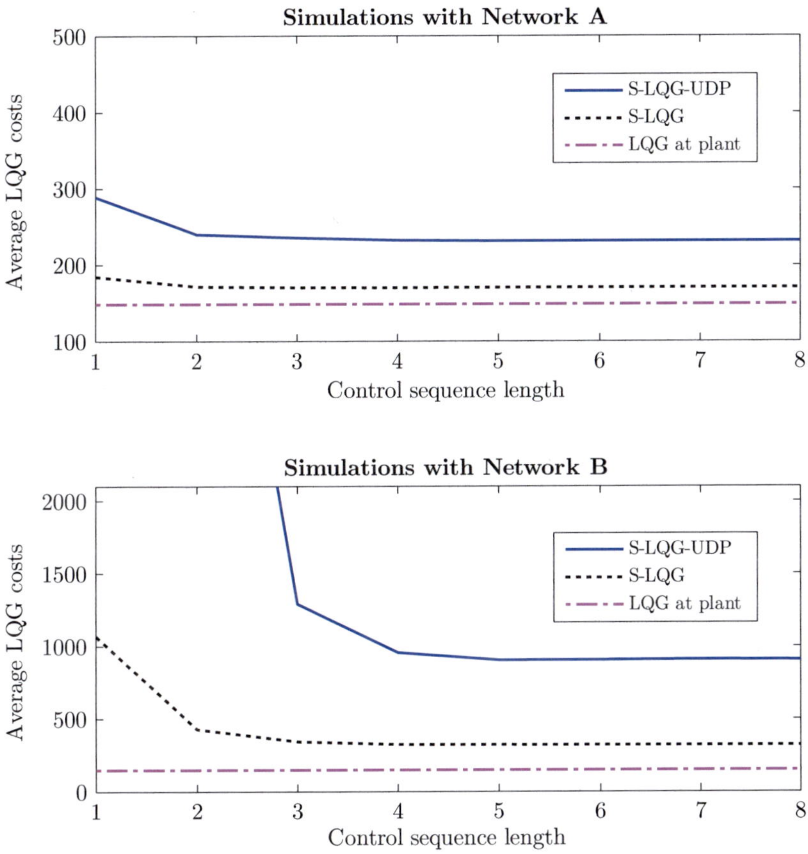

Figure 3.11.: Simulation results for certainty equivalence based approach

4. Sequence-Based Trajectory Tracking

The basic S-LQG controller stabilizes a system around the origin of the state space, i.e., it keeps the state close to the point $\underline{x} = \underline{0}$. An extension of this case is determining how to stabilize the system around an equilibrium point that is not zero, i.e., around a state $\bar{\underline{x}} = \mathbf{A}\bar{\underline{x}}$ with $\bar{\underline{x}} \neq \underline{0}$. This problem can be solved in the same way as in standard LQG control [2] by adding an appropriate constant feedforward term to the output of the S-LQG. The feedforward term is chosen so that it shifts the operating point from the origin to the desired equilibrium point.

In this chapter, we go one step further and consider the problem of tracking control. In tracking control, the controller has to be designed such that the output of the closed-loop system follows a given reference trajectory. Hence, the aforementioned task of stabilization around an equilibrium point can be seen as a special case of tracking control. In general, the tracking problem is more challenging than the stabilization problem as the closed-loop system not only has to be stabilized, but also has to follow the given reference trajectory [127]. This is even more important in a networked scenario where time delays and transmission losses can occur in the network connections.

It is interesting to note that the vast majority of NCS control methods exclusively deal with the stabilization problem. An overview of this work has already been given in Section 2.3. Recent approaches that address the problem of tracking control in the presence of time delays and/or transmission losses are described in [34, 145, 148, 155]. Here, the controller is designed such that the tracking error dynamics are guaranteed to be input-to-state stable [148] or the tracking error is minimized with respect to the H_∞-norm [34, 145] and the H_2-norm [155], respectively. However, the methods above only send a single control input per time step to the actuator and, hence, make no use of the sequence-based method.

As demonstrated in the previous chapter, the sequence-based control approach promises improved performance due to the active compensation of network-induced time delays and packet losses. Moreover, in the context of tracking control, the sequence-based method offers the additional advantage of embedding available information of the future reference trajectory in the out-going control sequences. This can be extremely valuable in situations such as robot control, where the reference trajectories are calculated in advance by path planning algorithms. The available preview information can be integrated into the control sequences sent to the actuator such that the plant still follows the reference trajectory, even if time delays and/or losses occur in the communication.

The tracking control problem in sequence-based control is explicitly addressed in the works [11, 70, 97, 134]. Here, the approaches [11, 134] use the sequence-based MPC approach described in Section 2.4.3, while [70, 97] are based on the extension of a nominal controller as discussed in Section 2.4.2. A stochastic optimal tracking controller for NCS with time-varying transmission delays and/or packet losses has not yet been proposed. The available optimal control approaches are only concerned with the stabilization problem and not with the tracking problem. For an overview of these non-tracking methods, the reader is referred to Section 2.4.4, where the general idea of the optimal sequence-based control is described.

In this chapter, we derive the sequence-based closed-loop optimal tracking controller for the networked LQG setup described in Section 3.1. Again, we make the assumption that the controller-actuator network provides idealized acknowledgments (Assumption 3.5). The approach was first published in [173] (own publication). The derived controller optimally compensates for time delays and transmission losses in the network connections and optimally incorporates preview information on the reference trajectory. An important result is that, similar to standard LQG control, the optimal controller can be separated without loss of optimality into a feedback part and a feedforward part. The feedback part is identical to the S-LQG controller. As we will see, this allows for easy application of the controller and fast adaption to online changes of the reference trajectory. The findings extend the results obtained in [159, 160], where this kind of separation is proven to hold for non-sequence-based tracking control over quantized networks.

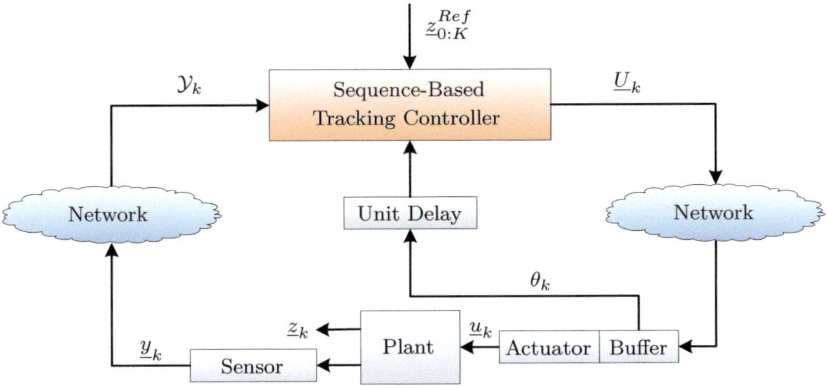

Figure 4.1.: NCS setup for sequence-based tracking control

In the next section, we will give a formal description of the tracking control problem. The optimal solution to this problem is presented in Section 4.2 and evaluated with simulations in Section 4.3. The formal derivation of the control law is given in Appendix C.

4.1. Problem Formulation

We consider the problem setup shown in Fig. 4.1. It is almost identical to the S-LQG setup described in Section 3.1. Yet, we add the plant output $\underline{z}_k \in \mathbb{R}^{n_z}$ that is considered for the tracking task. Thus, the system equations are given by

$$\begin{aligned} \underline{x}_{k+1} &= \mathbf{A}_k \underline{x}_k + \mathbf{B}_k \underline{u}_k + \underline{w}_k \ , \\ \underline{y}_k &= \mathbf{C}_k \underline{x}_k + \underline{v}_k \ , \\ \underline{z}_k &= \mathbf{Z}_k \underline{x}_k \ . \end{aligned} \tag{4.1}$$

The controller shall be designed such that $\underline{z}_k$ follows a given reference trajectory $\underline{z}_{0:K}^{Ref}$ with terminal time $K \in \mathbb{N}_{>0}$. For the sake of brevity, we assume that $\underline{z}_{0:K}^{Ref}$ is fixed and directly available to the controller. The setup can easily be extended to a case in which the reference trajectory may change during operation, which is discussed at the end of the next section.

We define the tracking error $\underline{\Delta}_k$ as the difference between the reference value $\underline{z}_k^{Ref}$ and the corresponding plant output $\underline{z}_k$ at time k, i.e.,

$$\underline{\Delta}_k \stackrel{\text{def}}{=} \underline{z}_k - \underline{z}_k^{Ref} \ .$$

To measure the tracking performance, we use the quadratic tracking error along with the energy consumed by the control. Both are brought together in the following finite-horizon cumulative cost function

$$C_{0\to K}^{Trk}(\underline{U}_{0:K}) \stackrel{\text{def}}{=} \mathrm{E}\left\{\underline{\Delta}_K^\top \mathbf{Q}_K \underline{\Delta}_K + \sum_{k=0}^{K-1} \left[\underline{\Delta}_k^\top \mathbf{Q}_k \underline{\Delta}_k + \underline{u}_k^\top \mathbf{R}_k \underline{u}_k\right] \middle| \mathcal{I}_0^{Trk}\right\} \ . \tag{4.2}$$

The weighting matrices $\mathbf{Q}_k \in \mathbb{R}^{n_z \times n_z}$ and $\mathbf{R}_k \in \mathbb{R}^{n_u \times n_u}$ are positive semidefinite and positive definite, respectively. Further, the information set $\mathcal{I}_k^{Trk}$, available to the controller at time step k, consists of the information set $\mathcal{I}_k$ (available to the S-LQG controller in Problem 3.1) extended by the given reference trajectory such that

$$\begin{aligned}\mathcal{I}_k^{Trk} &\stackrel{\text{def}}{=} \mathcal{I}_k \cup \left\{\underline{z}_{0:K}^{Ref}\right\} \\ &= \begin{cases} \left\{\underline{\bar{x}}_0, \mathbf{\Lambda}_0, \underline{z}_{0:K}^{Ref}\right\} & \text{for } k = 0 \ , \\ \left\{\mathcal{Y}_{1:k}, \underline{U}_{0:k-1}, \theta_{0:k-1}, \underline{\bar{x}}_0, \mathbf{\Lambda}_0, \underline{z}_{0:K}^{Ref}\right\} & \text{for } k \in \mathbb{N}_{>0} \ . \end{cases}\end{aligned} \tag{4.3}$$

Finally, we formalize the optimal sequence-based tracking control problem.

Problem 4.1 Sequence-Based LQG Tracking Control Problem (with zero-input strategy)

$$\min_{\underline{U}_{0:K-1}} C_{0\to K}^{Trk}\left(\underline{U}_{0:K-1}\right) \ ,$$

subject to: $\underline{U}_k = \mu_k\left(\mathcal{I}_k^{Trk}\right) \ , \quad \underline{u}_k^{df} = \underline{0} \ , \quad (4.1), \ (4.3), \ (3.10)\text{–}(3.11) \ .$

The solution to this optimization problem is presented in the next section. The corresponding proof can be found in Appendix C.

4.2. Optimal Control Law

The optimal control law that solves Problem 4.1 is summarized in the following theorem. In the exposition, we use the sequence-based NCS model defined by (3.29) that contains the augmented matrices $\widehat{\mathbf{A}}_k(\theta_k)$, $\widehat{\mathbf{B}}_k(\theta_k)$, and $\widehat{\mathbf{R}}_k(\theta_k)$ and the variable θ_k to specify the age of the sequence buffered in the actuator.

Theorem 4.1 *Consider the sequence-based LQG tracking problem 4.1 of minimizing* $C^{Trk}_{0\to K}\left(\underline{U}_{0:K-1}\right)$. *Then,*

a) the optimal control sequences can be separated into a feedback and a feedforward term such that

$$\underline{U}_k = \underline{U}_k^{fb} + \underline{U}_k^{ff} \ , \tag{4.4}$$

where the feedback term $\underline{U}_k^{fb}$ *depends on the estimated state of the system but not on the reference trajectory. The feedforward term* $\underline{U}_k^{ff}$ *depends on the reference trajectory but not on the system state.*

b) the control law to calculate the feedback term $\underline{U}_k^{fb}$ *is given by the S-LQG controller described in Theorem 3.1 , i.e.,*

$$\underline{U}_k^{fb} = \mathbf{L}_k(\theta_{k-1}) \cdot \mathrm{E}\left\{\underline{\xi}_k \middle| \mathcal{I}_k\right\} \ , \tag{4.5}$$

where $\mathbf{L}_k(\theta_{k-1})$ *is defined in (3.31).*

c) the control law to calculate the feedforward term $\underline{U}_k^{ff}$ *is given by*

$$\begin{aligned}\underline{U}_k^{ff} = &- \left(\mathrm{E}\left\{\widehat{\mathbf{R}}_k(\theta_k) + \widehat{\mathbf{B}}_k(\theta_k)^\top \mathbf{K}_{k+1} \widehat{\mathbf{B}}_k(\theta_k) \middle| \mathcal{I}_k\right\}\right)^\dagger \\ &\cdot \mathrm{E}\left\{\widehat{\mathbf{B}}_k(\theta_k)^\top \underline{\sigma}_{k+1} \middle| \mathcal{I}_k\right\},\end{aligned} \tag{4.6}$$

with

$$\mathbf{K}_k = \mathrm{E}\left\{\widehat{\mathbf{Q}}_k(\theta_k) + \widehat{\mathbf{A}}_k(\theta_k)^\top \mathbf{K}_{k+1} \widehat{\mathbf{A}}_k(\theta_k) \middle| \mathcal{I}_k\right\} - \mathbf{P}_k \ , \tag{4.7}$$

$$\begin{aligned} \mathbf{P}_k \stackrel{\text{def}}{=} & \mathrm{E}\left\{\widehat{\mathbf{A}}_k(\theta_k)^\top \mathbf{K}_{k+1} \widehat{\mathbf{B}}_k(\theta_k) \middle| \mathcal{I}_k\right\} \\ & \cdot \left(\mathrm{E}\left\{\widehat{\mathbf{R}}_k(\theta_k) + \widehat{\mathbf{B}}_k(\theta_k)^\top \mathbf{K}_{k+1} \widehat{\mathbf{B}}_k(\theta_k) \middle| \mathcal{I}_k\right\}\right)^\dagger \\ & \cdot \ \mathrm{E}\left\{\widehat{\mathbf{B}}_k(\theta_k)^\top \mathbf{K}_{k+1} \widehat{\mathbf{A}}_k(\theta_k) \middle| \mathcal{I}_k\right\} \ , \end{aligned} \tag{4.8}$$

$$\begin{aligned} \underline{\sigma}_k = & \mathrm{E}\left\{\widehat{\mathbf{A}}_k(\theta_k)^\top \underline{\sigma}_{k+1} \middle| \mathcal{I}_k\right\} + \overline{\mathbf{Q}}_k^\top \underline{z}_k^{Ref} \\ & - \mathrm{E}\left\{\widehat{\mathbf{A}}_k(\theta_k)^\top \mathbf{K}_{k+1} \widehat{\mathbf{B}}_k(\theta_k) \middle| \mathcal{I}_k\right\} \\ & \cdot \left(\mathrm{E}\left\{\widehat{\mathbf{R}}_k(\theta_k) + \widehat{\mathbf{B}}_k(\theta_k)^\top \mathbf{K}_{k+1} \widehat{\mathbf{B}}_k(\theta_k) \middle| \mathcal{I}_k\right\}\right)^\dagger \\ & \cdot \mathrm{E}\left\{\widehat{\mathbf{B}}_k(\theta_k)^\top \underline{\sigma}_{k+1} \middle| \mathcal{I}_k\right\} \ , \end{aligned} \tag{4.9}$$

and

$$\begin{aligned} & \widehat{\mathbf{R}}_k(\theta_k) \stackrel{\text{def}}{=} \mathbf{J}_k(\theta_k)^\top \mathbf{R}_k \mathbf{J}_k(\theta_k) \ , \qquad \overline{\mathbf{Q}}_k \stackrel{\text{def}}{=} \begin{bmatrix} \mathbf{Q}_k \mathbf{Z}_k & \mathbf{0} \end{bmatrix} \ , \\ & \widehat{\mathbf{Q}}_k(\theta_k) \stackrel{\text{def}}{=} \begin{bmatrix} \mathbf{Z}_k^\top \mathbf{Q}_k \mathbf{Z}_k & \mathbf{0} \\ \mathbf{0} & \mathbf{H}_k(\theta_k)^\top \mathbf{R}_k \mathbf{H}_k(\theta_k) \end{bmatrix} \ . \end{aligned} \tag{4.10}$$

The initial conditions are given by

$$\begin{aligned} & \mathbf{K}_K \stackrel{\text{def}}{=} \begin{bmatrix} \mathbf{Z}_K^\top \mathbf{Q}_K \mathbf{Z}_K & \mathbf{0} \\ \mathbf{0} & \mathbf{0} \end{bmatrix} \ , \quad \underline{\sigma}_K \stackrel{\text{def}}{=} \begin{bmatrix} \mathbf{Q}_K \mathbf{Z}_K & \mathbf{0} \end{bmatrix}^\top \underline{z}_K^{Ref} \ , \\ & s_K \stackrel{\text{def}}{=} (\underline{z}_K^{Ref})^\top \mathbf{Q}_K \underline{z}_K^{Ref} \ . \end{aligned} \tag{4.11}$$

d) the minimum expected cumulative costs are given by

$$\begin{aligned} C_{0\to K}^{Trk^*} \stackrel{\text{def}}{=} & \min_{\underline{U}_{0:K-1}} C_{0\to K}^{Trk}(\underline{U}_{0:K-1}) \\ & = \mathrm{E}\left\{\underline{\xi}_0^\top \mathbf{K}_0 \underline{\xi}_0 \middle| \mathcal{I}_0\right\} + \sum_{k=0}^{K-1} \mathrm{E}\left\{\underline{e}_k^\top \mathbf{P}_k \underline{e}_k \middle| \mathcal{I}_0\right\} + s_0 \\ & \quad - 2 \cdot \underline{\sigma}_0^\top \mathrm{E}\left\{\underline{\xi}_0 \middle| \mathcal{I}_0\right\} + \sum_{k=0}^{K-1} \mathrm{E}\left\{\underline{\widehat{w}}_k^\top \mathbf{K}_{k+1} \underline{\widehat{w}}_k \middle| \mathcal{I}_0\right\} \ , \end{aligned}$$

with

$$\underline{e}_k \stackrel{\text{def}}{=} \underline{\xi}_k - \mathrm{E}\left\{\underline{\xi}_k \middle| \mathcal{I}_k\right\} , \tag{4.12}$$

$$\begin{aligned} s_k &= \mathrm{E}\left\{s_{k+1} \middle| \mathcal{I}_k\right\} + (\underline{z}_k^{Ref})^\top \mathbf{Q}_k \underline{z}_k^{Ref} - \mathrm{E}\left\{\underline{\sigma}_{k+1}^\top \widehat{\mathbf{B}}_k(\theta_k) \middle| \mathcal{I}_k\right\} \\ &\cdot \left(\mathrm{E}\left\{\widehat{\mathbf{R}}_k(\theta_k) + \widehat{\mathbf{B}}_k(\theta_k)^\top \mathbf{K}_{k+1} \widehat{\mathbf{B}}_k(\theta_k) \middle| \mathcal{I}_k\right\}\right)^\dagger \\ &\cdot \mathrm{E}\left\{\widehat{\mathbf{B}}_k(\theta_k)^\top \underline{\sigma}_{k+1} \middle| \mathcal{I}_k\right\} . \end{aligned} \tag{4.13}$$

PROOF. The proof is given in Appendix C. □

The structure of the optimal controller described by Theorem 4.1 is illustrated in Fig. 4.2. Analogous to standard LQG tracking control, the resulting controller consists of two parts: a feedback term and a feedforward term. As can be seen by comparison of (4.5) with Theorem 3.1 c), the feedback term is identical to the finite horizon S-LQG controller. In this tracking scenario, the S-LQG has the purpose of attenuating stochastic disturbances by state feedback and stabilizing the system around the origin of the state space. The feedforward term then shifts the system state from the origin to the reference trajectory. With respect to the system state, the feedforward part is an open-loop control law. This can be seen by (4.6) as it only depends on the reference trajectory via (4.9). However, as the feedforward part (as well as the feedback part) also depends on the acknowledgment signals, it is not purely an open-loop control law.

With the reference trajectory known in advance, the matrices (4.6)–(4.11) can be precomputed. Hence, the control law of the optimal tracking controller can also be calculated offline. During operation, only the augmented state estimate has to be obtained (as discussed in Section 3.4.1) and applied to the corresponding precomputed matrices dependent on the current time index and acknowledgment signal. If the reference trajectory is only partially available or changes during operation, only the feedforward part (4.6) of the tracking controller has to be recomputed online. The feedback control law does not change.

If long time horizons have to be considered where changes of the reference trajectory are likely, it is also possible to apply the tracking controller in

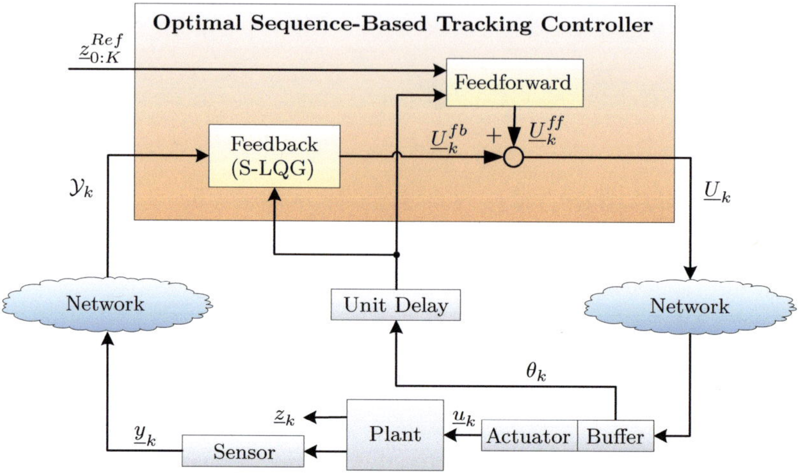

Figure 4.2.: Structure of the optimal sequence-based tracking controller

a receding horizon scheme as in sequence-based MPC. In this operation mode, the reference trajectory is periodically recalculated over a finite time horizon shorter than the horizon of the tracking task. This scheme even allows for infinite-horizon tracking control. An appealing advantage of considering long time horizons is that the feedback gain of the underlying S-LQG converges (under assumptions) to a steady state as stated in Theorem 3.1 . This significantly reduces the memory required for the tracking controller.

4.3. Evaluation

In the following, we illustrate the behavior of the optimal tracking controller by simulations with an inverted pendulum on a cart that is controlled over a network. The inverted pendulum is modeled as described in [3]. The state of the pendulum is given by

$$\underline{x}(t) = \begin{bmatrix} s(t) & \dot{s}(t) & \phi(t) & \dot{\phi}(t) \end{bmatrix}^\top ,$$

Mass of the cart	0.5 kg
Mass of the pendulum	0.5 kg
Friction of the cart	0.1 N/(ms)
Length to pendulum center of mass	0.3 m
Inertia of the pendulum	0.006 kgm^2
Sampling time	0.1 s

Table 4.1.: Parameters of inverted pendulum used in simulations

where $s(t)$ is the position of the cart and $\phi(t)$ is the angle of the pendulum rod relative to an upright position. We consider a pendulum with parameters as shown in Table 4.1. The discrete-time state space model[1] of this system is then given by

$$\mathbf{A} = \begin{bmatrix} 1.0000 & 0.0200 & 0.0015 & 0.0000 \\ 0 & 0.9964 & 0.1550 & 0.0015 \\ 0 & -0.0001 & 1.0103 & 0.0201 \\ 0 & -0.0105 & 0.0343 & 1.0103 \end{bmatrix}, \quad \mathbf{B} = \begin{bmatrix} 0.0004 \\ 0.0358 \\ 0.0011 \\ 0.1054 \end{bmatrix},$$

$$\mathbf{C} = \begin{bmatrix} 1 & 0 & 0 & 0 \\ 0 & 0 & 1 & 0 \end{bmatrix}.$$

In the simulation, we choose the covariances of the disturbances, the initial condition, and the weighting matrices to

$$\mathbf{W} = \mathbf{V} = \begin{bmatrix} 0.005^2 & 0 & 0 & 0 \\ 0 & 0 & 0 & 0 \\ 0 & 0 & \left(\frac{0.2\cdot\pi}{360}\right)^2 & 0 \\ 0 & 0 & 0 & 0 \end{bmatrix},$$

$$\underline{\bar{x}}_0 = \begin{bmatrix} 0 \\ 0.2 \\ 0 \\ 0.2 \end{bmatrix}, \quad \mathbf{\Lambda}_0 = \begin{bmatrix} 0.01^2 & 0 & 0 & 0 \\ 0 & 0 & 0 & 0 \\ 0 & 0 & 0.01^2 & 0 \\ 0 & 0 & 0 & 0 \end{bmatrix}, \quad \mathbf{Q} = \mathbf{R} = \mathbf{Z} = \mathbf{I}.$$

The networks are modeled as two independent data links with stochastic characteristics as shown in Fig. 4.3.

[1] Units are are omitted in the model for reasons of clarity.

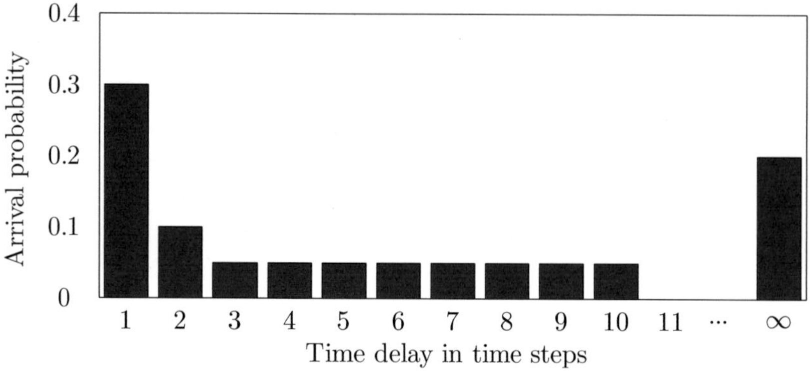

Figure 4.3.: Probabilities of transmission delays in the network connections used in the simulation. Transmission losses are considered as infinite time delays.

We calculate the optimal tracking controller for different control sequence lengths where the default control input is set to $\underline{u}^{df} = \underline{0}$. The reference trajectory is shown in Fig. 4.4. For comparison to another sequence-based controller, we implement the approach described in [70] that is based on a nominal controller The nominal controller is implemented as an optimal linear quadratic tracking controller [2]. At the controller site, the state is estimated using the optimal estimator derived in [81], which is a time-varying Kalman filter with measurement buffer.

A typical sample run for each controller tracking the reference trajectory is shown in Fig. 4.4. The length of the control sequence is $N = 4$. At first, the trajectories are almost identical and the cart moves from the initial position $s(0)$ to the reference value of 50. Then, however, the proposed optimal tracking controller already orientates after four seconds towards the new reference value of -50, although this reference will not be active for another second. The extended nominal controller does not anticipate the set point change. As a result, higher fluctuations of angle and angular velocity occur than with the proposed optimal sequence-based controller with preview.

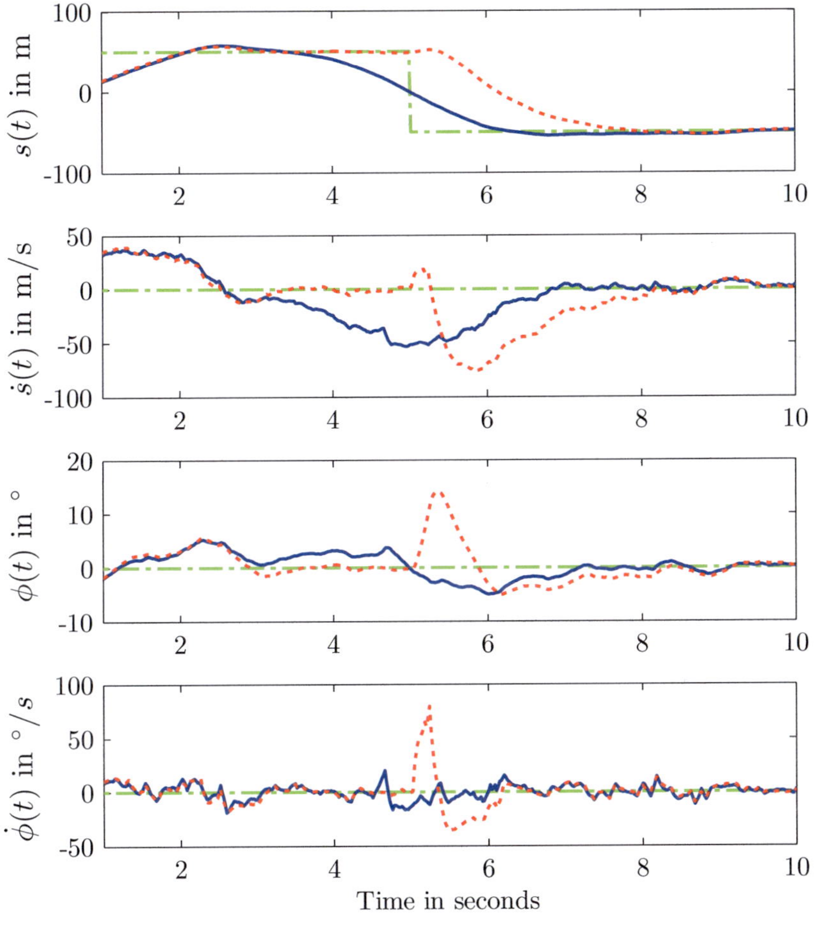

Figure 4.4.: Example of state trajectory for a simulation with control sequence length $N = 4$ for proposed controller (—) and control approach of [70] (– –) when tracking the reference trajectory (– · –). The position and velocities of the cart and the rod are shown.

This also has an influence on the resulting costs. Performing 100 Monte Carlo simulation runs over 500 time steps each, the average costs are calculated according to (4.2) over all simulation runs. The outcome is shown in Fig. 4.5 for different lengths of the control sequence. We notice that the average costs of both controllers decrease with increasing sequence length. This underpins the benefit of the sequence-based control method. As a second observation, we can state that the proposed tracking controller indeed leads to lower costs than the nominal-controller-based approach. The latter is even unstable for sequence lengths shorter than $N = 3$.

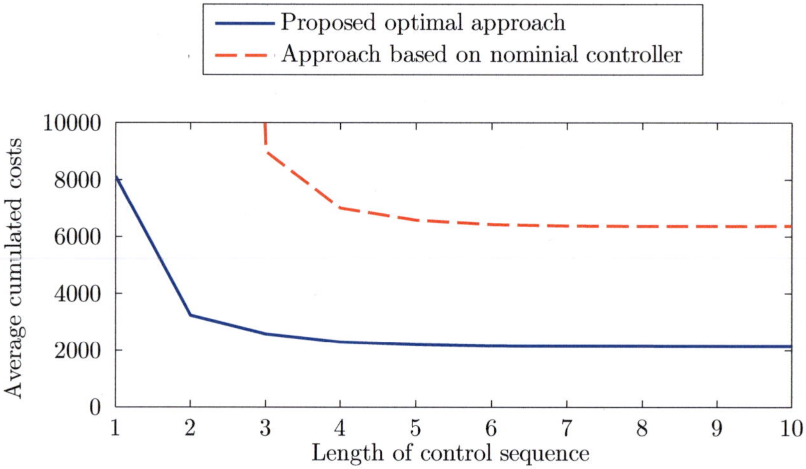

Figure 4.5.: Comparison of the cumulative average costs (4.2) over different lengths of the control sequence for 100 Monte Carlo simulation runs.

Concluding this section, we saw in the example that despite packet losses and time-varying transmission delays, preview information on the reference trajectory can be profitably used to increase tracking performance. Thus, the incorporation of reference preview within the sequence-based control framework is a natural extension of the general sequence-based control method. Furthermore, these benefits do not come at additional communication costs.

5. Event-Triggered S-LQG Control

The S-LQG is designed to compensate for stochastic packet losses and time-varying transmission delays in NCS. As described in Section 2.2.3, band-limitations are further challenges in control over digital networks. In particular, these can restrict the number of possible data transmissions between sensors, controllers, and actuators. Therefore, it is desirable to send data over the network only when necessary. This also makes sense under consideration of a possible deterioration of the network due to high network load. As described in Section 2.2, transmitting data increases the network load and, hence, can significantly increase time delays and packet losses. Therefore, a lot of research has been conducted to develop control schemes that minimize the communication expenditure. The proposed approaches can be distinguished into two classes. The first one is concerned with the problem of finding optimal scheduling protocols that manage the network access on a MAC-layer level [144]. The other approaches are based on the idea of *event-triggered control* which is also referred to as *event-based control* [7, 24, 48, 64].

The main characteristic of event-triggered control is that data are only transmitted when a certain event occurs. Measurements could, for example, only be sent from the sensor to the controller if the difference between the last transmitted measurement and the current measurement is larger than a certain threshold, or the estimation error covariance exceeds a specified bound [35, 84, 85, 139]. In the same way, the controller can decide not to send a control input to the actuator if the current control input already provides a sufficient control performance [48]. In this scheme, the control input is held at the actuator until a new admissible packet is received. A recent survey of event-triggered methods can be found in [74] and recent work concerning the non-sequence-based setting is given in [5, 36, 64, 78, 143].

In this chapter, an approach is presented that combines the idea of event-triggered control with sequence-based control. The results are published in abbreviated form in [170] (own publication). The proposed controller is based on the S-LQG and not only compensates for time-varying transmission delays and packet losses in the controller-actuator network, but also reduces the communication. Data transmissions are only initiated if this is justified by sufficient improvement to the system performance. The combination of these two methods is extremely beneficial as the control sequences buffered in the actuator may guarantee a sufficient control performance for quite a long time. Therefore, the number of data transmissions can be greatly decreased. Of course, the sequence-based controller sends more information per data packet than a controller that only computes single control inputs. However, assuming that the control sequences are optimally used at the actuator site, we can expect an effective reduction of the data transmitted due to less overhead. For example, the Internet protocol IPv4 typically has an overhead of 20 to 60 bytes per data packet whereas a control input often only occupies around 2 bytes. Therefore, sending fewer data packets with higher payload reduces the overhead so that the total amount of data communicated can be reduced [37, 165].

Combined sequence-based and event-triggered control has been considered previously. A sequence-based LQ optimization approach is described in [40]. The authors derive a sequence-based self-triggered LQR controller that calculates the transmission times. However, the approach does not incorporate network-induced delays or losses and requires perfect state information. In addition, the approaches in [165, 168] consider an event-triggered sequence-based control using the generalized nominal controller approach described in Section 2.4.2. In [168], the controller sends a control sequence as soon as the difference between the buffered control sequence and the newly calculated control sequence exceeds a certain threshold. The difference of the two sequences is determined based on the maximum norm. In [165], the event-triggered sensor only sends a measurement to the controller if the measurement significantly differs from the last one sent. As soon as the controller receives a triggered measurement, a control sequence is calculated and sent to the actuator. In contrast to the approach presented in this chapter, the decision rule in [165, 168] is not based on the expected costs, but on empirical tuning parameters.

In the next section, the this problem setup is introduced in detail. Then, the proposed controller is presented in Section 5.2 and the stability of the closed-loop is analyzed in Section 5.3. Finally, the performance is evaluated in Monte Carlo simulations with an inverted pendulum in Section 5.4.

5.1. Problem Formulation

Here, we consider the S-LQG system setup introduced in Section 3.1 where system performance is measured in terms of the quadratic cost function (3.16). For convenience, the cost function is restated here

$$C_{0\to K}(\underline{U}_{0:K-1}) = \mathop{\mathrm{E}}_{\substack{\underline{x}_{0:K}\\ \underline{u}_{0:K-1}}}\left\{\underline{x}_K^\top \mathbf{Q}_K \underline{x}_K + \sum_{k=0}^{K-1} \underline{x}_k^\top \mathbf{Q}_k \underline{x}_k + \underline{u}_k^\top \mathbf{R}_k \underline{u}_k \,\middle|\, \mathcal{I}_0, \underline{U}_{0:K-1}\right\}.$$

One way to address the problem of simultaneously minimizing the cost function and reducing the communication between the controller and the actuator is to extend the cost function by an additional term that represents the communication costs [74]. Therefore, we introduce the extended cost function

$$\begin{aligned} C_{0\to K}^{Evt}(\underline{U}_{0:K-1}, s_{0:K-1}) \stackrel{\text{def}}{=} \mathop{\mathrm{E}}_{\substack{\underline{x}_{0:K}\\ \underline{u}_{0:K-1}}}\Bigg\{&\underline{x}_K^\top \mathbf{Q}_K \underline{x}_K \\ &+ \sum_{k=0}^{K-1} \left(S_k s_k + \underline{x}_k^\top \mathbf{Q}_k \underline{x}_k + \underline{u}_k^\top \mathbf{R}_k \underline{u}_k\right) \Bigg| \mathcal{I}_0, \underline{U}_{0:K-1}, s_{0:K-1}\Bigg\} . \end{aligned} \tag{5.1}$$

This cost function contains the additional term $S_k s_k$ where S_k is a positive scalar and $s_k \in \{0, 1\}$. The scalar S_k represents the costs for one data transmission while s_k determines whether the current control sequence $\underline{U}_k$ shall be sent or not, i.e.,

$$s_k = \begin{cases} 1 & \text{if } \underline{U}_k \text{ is sent to actuator ,} \\ 0 & \text{if } \underline{U}_k \text{ is not sent to actuator .} \end{cases} \tag{5.2}$$

The transmission costs S_k only occur in the cumulative costs if the controller sends the data packet $\underline{U}_k$ to the actuator. If the control sequence is not sent, no transmission costs are incurred. Analogous to the weighting matrices $\mathbf{Q}_k$ and $\mathbf{R}_k$, the transmission cost parameter S_k is a design parameter and has to be chosen such that desired system specifications are met.

At each time step, the event-triggered controller not only calculates $\underline{U}_k$, but also the decision variable s_k. Depending on s_k, the sequence $\underline{U}_k$ is sent or not. We denote the control law of the event-triggered controller by the set of functions $\{\mu_0^{Evt}(\cdot), \ldots, \mu_K^{Evt}(\cdot)\}$. Deviating from Definition 3.2, we call a control law admissible if it holds for all $k \in \mathbb{N}_0$

$$\{\underline{U}_k, s_k\} = \mu_k^{Evt}\left(\mathcal{I}_k^{Evt}\right) , \tag{5.3}$$

where $\mathcal{I}_k^{Evt}$ denotes the information available to the event-triggered controller at time step k. The information set $\mathcal{I}_k^{Evt}$ comprises the S-LQG information set $\mathcal{I}_k$ defined in (3.13) and the past sending decisions $s_{0:k-1}$ such that

$$\begin{aligned} \mathcal{I}_k^{Evt} &\stackrel{\text{def}}{=} \mathcal{I}_k \cup \{s_{0:k-1}\} \\ &= \begin{cases} \{\underline{x}_0, \mathbf{\Lambda}_0\} & \text{for } \ k = 0 \ , \\ \{s_{0:k-1}, \mathcal{Y}_{1:k}, \underline{U}_{0:k-1}, \theta_{0:k-1}, \underline{x}_0, \mathbf{\Lambda}_0\} & \text{for } \ k \in \mathbb{N}_{>0} \ . \end{cases} \end{aligned} \tag{5.4}$$

With these definitions, we are able to formalize the event-triggered sequence-based control problem.

Problem 5.1 Event-Triggered Sequence-Based LQG Control

$$\min_{\substack{\underline{U}_{0:K-1} \\ s_{0:K-1}}} C_{0\to K}^{Evt}\left(\underline{U}_{0:K-1}, s_{0:K-1}\right)$$

$$\text{subject to:} \quad \{\underline{U}_k, s_k\} = \mu_k^{Evt}\left(\mathcal{I}_k^{Evt}\right), (4.1),\ (4.3), \text{and } (3.10) - (3.11) \ .$$

The optimization problem contains the discrete-valued decision variables $s_{0:K-1}$ and the continuous-valued decision variables $\underline{U}_{0:K-1}$. Hence, this is a hybrid optimization problem. It is generally recognized that these kind of optimization problems are extremely hard to solve optimally as the decision tree spanned by the discrete variable grows exponentially with the length of the time horizon. Thus, we do attend to find the optimal controller, but rather propose a reasonable approximate solution in the next section.

5.2. Proposed Solution

We propose an approximate solution to the event-triggered sequence-based S-LQG control problem 5.1 that is based on a so called rollout strategy. To that end, we separate the control algorithm $\mu_k^{Evt}(\mathcal{I}_k^{Evt})$ at each time step into the following two steps:

1. The first step consists of computing an optimal control sequence candidate $\underline{U}_k^{cand}$ that is eligible to be sent to the actuator at time step k.

2. In the second step, the controller determines the minimum expected costs for following two cases: a) the control sequence candidate is sent to the actuator and b) the control sequence candidate is not sent to the actuator. If sending leads to lower expected costs, the control sequence candidate is transmitted, otherwise it is discarded and no sequence is sent at this time step.

Note that the separation of the control law into these two steps is not yet an approximation. For non-trivial time horizons, however, there is neither a known solution to analytically compute the optimal control sequence candidate, nor to determine the minimum expected costs. Therefore, we will apply the following rollout strategy that allows us to use a modified version of the S-LQG to calculate a control sequence candidate and to evaluate the resulting expected costs.

Approximation 5.1 To calculate the control sequence candidate and the minimum expected costs at a certain time step, we make the (most likely wrong) assumption that all future control sequences will be sent to the actuator. This means that we assume in the calculations of $\underline{U}_k^{cand}$ and costs (5.1) that $s_{k+1:K} = 1$.

Of course, this strategy is an approximation, as the controller could decide not to send a control sequence at any subsequent time step. In context of dynamic programming, this kind of approximation is called a rollout strategy, as it approximates the expected cost-to-go of future time steps [15]. Therefore, we will refer to Approximation 5.1 also as rollout strategy.

By applying the rollout strategy, the aforementioned steps of the control algorithm can be evaluated. The structure of the resulting controller is shown in Fig. 5.1. In the following, we describe both the modified S-LQG to calculate the control sequence candidate $\underline{U}_k$ and the event-triggered decision rule.

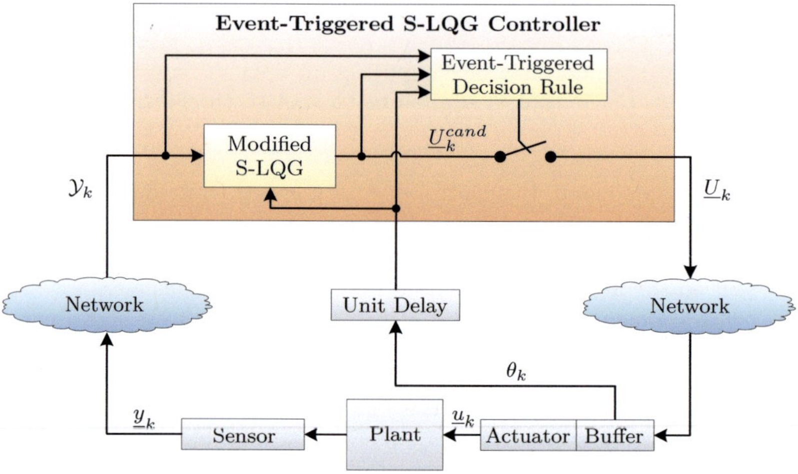

Figure 5.1.: Structure of the proposed event-triggered controller

5.2.1. Calculation of the Control Sequence Candidate

In this section, we will use Approximation 5.1 to calculate the control sequence candidate. To distinguish true decision variables s_k from believed decision variables resulting from application of Approximation 5.1 at time step k, we introduce the following definition.

Definition 5.1

$$s_l^{[k]} \stackrel{\text{def}}{=} \begin{cases} s_l & \text{for} \quad l \leq k \\ 1 & \text{for} \quad l > k \end{cases} \qquad \text{with } l, k \in \mathbb{N}_0 \text{ and } l \neq k \; . \tag{5.5}$$

The term $s_l^{[k]}$ can be interpreted as the believed value of s_l if Approximation 5.1 is applied at time step k. Based on this definition, we also introduce the information set

$$\mathcal{S}_l^{[k]} = \left\{ \left\{ s_k^{[k]} = 1 \right\}, s_{0:l-1}^{[k]}, \mathcal{Y}_{0:l}, \underline{U}_{0:l-1}, \theta_{0:l-1} \right\} \tag{5.6}$$

that describes the information available to the controller at time step l if Approximation 5.1 is applied at time step k. The information includes the tentative decision $s_k = 1$ (and hence $s_k^{[k]} = 1$), since we have to assume that the current sequence is sent. Otherwise, the calculation of a candidate sequence for time step k would be meaningless. Note that $\mathcal{S}_l^{[k]}$ is not necessarily a subset of $\mathcal{S}_l^{[k+m]}$ with $m \in \mathbb{N}_{>0}$ as it is assumed $s_{k:K}^{[k]} = 1$.

With these definitions, we can apply the rollout strategy to the original optimization problem 5.1 and calculate the approximated costs

$$C_{k\to K}^{Evt[k]} \stackrel{\text{def}}{=} \mathop{\mathrm{E}}_{\substack{\underline{x}_{k:K} \\ \underline{u}_{k:K-1}}} \left\{ \underline{x}_K^\top \mathbf{Q}_K \underline{x}_K + \sum_{l=k}^{K-1} \left(S_l s_l^{[k]} + \underline{x}_l^\top \mathbf{Q}_l \underline{x}_l + \underline{u}_l^\top \mathbf{R}_l \underline{u}_l \right) \middle| \mathcal{S}_l^{[k]} \right\}.$$

The control sequence candidate $\underline{U}_k^{cand}$ can be calculated by minimizing these costs over admissible control sequences under application of Approximation 5.1 according to

$$\begin{aligned} \underline{U}_{k:K-1}^{[k]} &\stackrel{\text{def}}{=} \mathop{\mathrm{argmin}}_{\substack{\underline{U}_k^{[k]} = \mu_k^{Evt[k]}\left(\mathcal{S}_k^{[k]}\right) \\ \vdots \\ \underline{U}_{K-1}^{[k]} = \mu_{K-1}^{Evt[k]}\left(\mathcal{S}_{K-1}^{[k]}\right)}} C_{k\to K}^{Evt[k]}, \\ \underline{U}_k^{cand} &= \underline{U}_k^{[k]}. \end{aligned} \tag{5.7}$$

Here, we have introduced the upper index $[k]$ for $\underline{U}_l^{[k]}$ and $\mu_k^{Evt[k]}(\cdot)$ to clarify that the control law $\mu_k^{Evt[k]}(\cdot), \ldots, \mu_{k+K-1}^{Evt[k]}(\cdot)$ for calculation of $\underline{U}_{k:K-1}$ is based on the application of Approximation 5.1 at time step k. This implies that a control law computed at time step k is only valid at this time step and in general cannot be used at future time instances. In particular, the tempting equalities $\mu_{k+1}^{Evt[k+1]}(\cdot) = \mu_{k+1}^{Evt[k]}(\cdot)$ and $\underline{U}_{k+1}^{[k+1]} = \underline{U}_{k+1}^{[k]}$ do not hold. Similarly to MPC, we therefore have to solve the optimization problem (5.7) again at each time step.

Remark 5.1 The optimization problem (5.7) seems to be an open-loop feedback (OLF) control problem [10]. Note, however, that the scheme presented is not an OLF approximation as it utilizes future feedback from measurements $\mathcal{Y}_{0:k}$ and acknowledgment signals.

In the following, we solve the optimization problem (5.7). To that end, we need to extend the system model of the S-LQG such that it captures the effects when a control sequence is not sent to the actuator. In this context, it is important to note that the decision of whether a sequence is sent or not directly influences the probabilities that certain sequences are buffered in the actuator. Therefore, the age of the buffered sequence θ_k and its evolution will depend on the decision variables $s_{0:k}$. Analyzing this dependency, it turns out that θ_k is only influenced by the last N sending decisions $s_{(k-N):k}$. Older decisions $s_{0:(k-N-1)}$ have no influence on the value of θ_k as the corresponding control sequences $\underline{U}_{0:(k-N-1)}$ do not contain control inputs applicable at time step k and, thus, can no longer be buffered in the actuator.

Consequently, the transition matrix of θ_k is a time-variant function depending on $s_{(k-N):k}$. This is the major difference compared to the system model used in the derivation of the S-LQG (see Section 3.3). In the following, we denote the time-variant transmission matrix of the event-triggered scheme by $\mathbf{T}_k^{Evt}$ and its entries by $p_k^{Evt}(i,j)$. The entries are calculated in a manner similar to Lemma 3.1 which was derived in the context of the standard S-LQG. However, the equations (3.22) and (3.23) have to be extended by $s_{(k-N):k}$ according to

$$p_k^{Evt}(i,j) = \begin{cases} 0 & \text{for } j \geq i+2 \, , \\ \prod_{m=0}^{i} \left(1 - s_{k-m} \cdot \check{q}^{CA}(m)\right) & \text{for } j = i+1 \, , \\ s_{k-j} \cdot \check{q}^{CA}(j) \prod_{m=0}^{j-1} \left(1 - s_{k-m} \cdot \check{q}^{CA}(m)\right) & \text{for } j < i \leq N \, , \\ \prod_{m=0}^{N-1} \left(1 - s_{k-m} \cdot \check{q}^{CA}(m)\right) & \text{for } j = i = N \, . \end{cases} \tag{5.8}$$

The term $\check{q}^{CA}(\cdot)$ is defined in (3.21). Since $\mathbf{T}_k^{Evt}$ depends on $s_{(k-Nk):k}$, there are 2^{N+1} possible transmission matrices, each corresponding to a unique history of the past N preceding control sequence candidates.

In analogy to s_l and $s_l^{[k]}$, we introduce the matrix $\mathbf{T}_l^{Evt[k]}$ to distinguish the accurate transition matrix $\mathbf{T}_l^{Evt}$ governing the evolution of θ_l from the approximated transition matrix $\mathbf{T}_l^{Evt[k]}$ that results from the application of Approximation 5.1 at time step k. The entries of the approximated transition matrix are denoted accordingly as $p_l^{Evt[k]}(i,j)$ and given by

$$p_l^{Evt[k]}(i,j) = \begin{cases} 0 & \text{for } j \geq i+2 \,, \\ \prod_{m=0}^{i} \left(1 - s_{l-m}^{[k]} \cdot \check{q}^{CA}(m)\right) & \text{for } j = i+1 \,, \\ s_{l-j}^{[k]} \cdot \check{q}^{CA}(j) \prod_{m=0}^{j-1} \left(1 - s_{l-m}^{[k]} \cdot \check{q}^{CA}(m)\right) & \text{for } j < i \leq N \,, \\ \prod_{m=0}^{N-1} \left(1 - s_{l-m}^{[k]} \cdot \check{q}^{CA}(m)\right) & \text{for } j = i = N \,. \end{cases} \tag{5.9}$$

With these definitions, we can finally formulate the solution of (5.7) in the following theorem.

Theorem 5.1 *The control sequence candidate $\underline{U}_k^{cand}$ that minimizes (5.7) is given by*

$$\underline{U}_k^{cand} = \mathbf{L}_k(\theta_{k-1}) \cdot \mathrm{E}\left\{\underline{\xi}_k \middle| \mathcal{I}_k\right\}, \tag{5.10}$$

where the feedback matrix $\mathbf{L}_k(\theta_{k-1})$ can be calculated for all realizations $\theta_{k-1} = i$ with $i \in \mathbb{J}$ by

$$\begin{aligned} \mathbf{L}_k(i) = &- \left(\sum_{j=0}^{N} p_k^{Evt}(i,j) \left[\widehat{\mathbf{R}}_k(j) + \widehat{\mathbf{B}}_k(j)^\top \mathbf{K}_{k+1}^{[k]}(j) \widehat{\mathbf{B}}_k(j) \right] \right)^\dagger \\ &\cdot \left(\sum_{j=0}^{N} p_k^{Evt}(i,j) \left[\widehat{\mathbf{B}}_k(j)^\top \mathbf{K}_{k+1}^{[k]}(j) \widehat{\mathbf{A}}_k(j) \right] \right) , \end{aligned} \tag{5.11}$$

and $\mathbf{K}_{k+1}^{[k]}(j)$ *is the solution of the recursion*

$$\begin{aligned}\mathbf{K}_l^{[k]}(i) = &\sum_{j=0}^{N} p_l^{Evt[k]}(i,j)\left[\widehat{\mathbf{Q}}_l(j) + \widehat{\mathbf{A}}_l(j)^\top \mathbf{K}_{l+1}^{[k]}(j)\widehat{\mathbf{A}}_l(j)\right] \\ &- \left(\sum_{j=0}^{N} p_l^{Evt[k]}(i,j)\left[\widehat{\mathbf{A}}_l(j)^\top \mathbf{K}_{l+1}^{[k]}(j)\widehat{\mathbf{B}}_l(j)\right]\right) \\ &\cdot \left(\sum_{j=0}^{N} p_l^{Evt[k]}(i,j)\left[\widehat{\mathbf{R}}_l(j) + \widehat{\mathbf{B}}_l(j)^\top \mathbf{K}_{l+1}^{[k]}(j)\widehat{\mathbf{B}}_l(j)\right]\right)^\dagger \\ &\cdot \left(\sum_{j=0}^{N} p_l^{Evt[k]}(i,j)\left[\widehat{\mathbf{B}}_l(j)^\top \mathbf{K}_{l+1}^{[k]}(j)\widehat{\mathbf{A}}_l(j)\right]\right),\end{aligned} \tag{5.12}$$

evolving backwards in time and initialized with $\mathbf{K}_K^{[k]}(i) = \begin{bmatrix}\mathbf{Q}_K & \mathbf{0}\\ \mathbf{0} & \mathbf{0}\end{bmatrix}$.

PROOF. The proof is analogous to the proof of Theorem 3.1 for the S-LQG with $p(i,j)$ and $\mathcal{I}_k$ replaced by $p_l^{Evt[k]}(i,j)$ and $\mathcal{S}_l^{[k]}$, respectively. □

As already mentioned, the optimization problem (5.7) has to be solved at each time step k. This is also reflected in the solution above by the presence of the upper index $[k]$ in several variables. For example, we see that the recursion (5.12) in Theorem 5.1 depends on the variable $p_l^{Evt[k]}(i,j)$. Therefore, $\mathbf{K}_l^{[k]}(j)$ explicitly depends on the time step at which Approximation 5.1 is made. As a result, the recursion (5.12) must theoretically be recomputed at each time step, since, in general, the matrix $\mathbf{K}_{k:K}^{[k]}(j)$ cannot be used to calculate the matrix $\mathbf{K}_{k+1:K}^{[k+1]}(j)$. This is in contrast to the S-LQG, where the recursion only has to be calculated once.

However, looking deeper into the evolution of θ_k, we can observe that the transition probabilities $p_l^{Evt[k]}(i,j)$ strictly converge towards the entries $p(i,j)$ of the S-LQG transition matrix given in Lemma 3.1. Indeed, the convergence is already completed after N time steps. This can be directly induced from (5.9) as it only depends on the N previous decisions $s_{(k-N):k}$. The fact greatly reduces the computational complexity because we can use the S-LQG to determine $\mathbf{K}^{[k]}_{k+N:K}(j)$ and then only have to perform the last N iterations of (5.12) during run time. We summarize this finding in the following corollary.

Corollary 5.1 The matrix $\mathbf{K}^{[k]}_{k+1}(j)$ needed in (5.11) to calculate the control sequence candidate $\underline{U}^{cand}_k$ can be computed by initializing the recursion (5.12) with the matrix $\mathbf{K}^{[k]}_{k+N}(i) = \mathbf{K}_k(i)$, where $\mathbf{K}_k(i)$ is the solution of the recursion (3.32) obtained in the derivation of the S-LQG.

In Section 5.3, we will use Corollary 5.1 to transfer stability and convergence properties of the S-LQG controller to the proposed event-triggered scheme. In particular, considering long time horizons, the S-LQG gain matrices converge (under assumptions). This implies that the control law in Theorem 5.1 can be computed offline even for infinite time horizons. Further aspects on this subject are discussed in Section 5.3.

5.2.2. Decision Step

After calculating the control sequence candidate, it is determined in the decision step whether the candidate sequence is sent to the actuator or not. The decision is made based on the expected costs that are incurred for each case. To evaluate these costs, we again make the assumption that all future control sequences are sent to the actuator (see Approximation 5.1). Recall the information set

$$\mathcal{S}^{[k]}_l = \left\{\left\{s^{[k]}_k = 1\right\}, s^{[k]}_{0:l-1}, \mathcal{Y}_{0:l}, \underline{U}_{0:l-1}, \theta_{0:l-1}\right\} ,$$

which is defined in (5.6). The set describes the information structure indicating that the calculated control sequence candidate $\underline{U}^{cand}_k$ is sent to the actuator. In analogy, we also define the set

$$\mathcal{N}^{[k]}_l \stackrel{\text{def}}{=} \left\{\left\{s^{[k]}_k = 0\right\}, s^{[k]}_{0:l-1}, \mathcal{Y}_{0:l}, \underline{U}_{0:l-1}, \theta_{0:l-1}\right\} , \tag{5.13}$$

that describes the information structure for the case in which the control sequence candidate is not sent. Based on these information sets, the minimum expected costs (5.1) achievable at time step k when using Approximation 5.1 can be analytically determined for each of the cases. The costs incurred by sending (or not sending) the candidate sequence are given by

$$C_{k\to K}^{Send} \stackrel{\text{def}}{=} \min_{\underline{U}_{k:K}} \mathrm{E}_{\substack{\underline{x}_{k:K}\\ \underline{u}_{k:K-1}}} \left\{ \underline{x}_K^\top \mathbf{Q}_K \underline{x}_K + \sum_{l=k}^{K-1} \left(S_l s_l + \underline{x}_l^\top \mathbf{Q}_l \underline{x}_l + \underline{u}_l^\top \mathbf{R}_l \underline{u}_l\right) \middle| \mathcal{S}_l^{[k]}, \underline{U}_{k:K-1} \right\}, \tag{5.14}$$

$$C_{k\to K}^{NotSend} \stackrel{\text{def}}{=} \min_{\underline{U}_{k:K}} \mathrm{E}_{\substack{\underline{x}_{k:K}\\ \underline{u}_{k:K-1}}} \left\{ \underline{x}_K^\top \mathbf{Q}_K \underline{x}_K + \sum_{l=k}^{K-1} \left(S_l s_l + \underline{x}_l^\top \mathbf{Q}_l \underline{x}_l + \underline{u}_l^\top \mathbf{R}_l \underline{u}_l\right) \middle| \mathcal{N}_l^{[k]}, \underline{U}_{k:K-1} \right\}. \tag{5.15}$$

The decision rule is defined based on the difference of these costs:

$$\begin{aligned} &\text{if} \quad C_{k\to K}^{Send} - C_{k\to K}^{NotSend} \geq 0\,, \quad \text{then do not send } \underline{U}_k^{cand} &&\Rightarrow \text{set } s_k = 0\,,\\ &\text{if} \quad C_{k\to K}^{Send} - C_{k\to K}^{NotSend} < 0\,, \quad \text{then send } \underline{U}_k^{cand} &&\Rightarrow \text{set } s_k = 1\,. \end{aligned} \tag{5.16}$$

The difference $C_{k\to K}^{Send} - C_{k\to K}^{NotSend}$ can be calculated in closed form following the applied rollout strategy. The obtained result is summarized in the following lemma.

Lemma 5.1 The term $C_{k\to K}^{Send} - C_{k\to K}^{NotSend}$ that describes the difference between the minimum expected costs (5.1) under Approximation 5.1 of sending and not sending the candidate sequence is obtained by

$$\begin{aligned} C_{k\to K}^{Send} - C_{k\to K}^{NotSend} &= \mathrm{E}\left\{\underline{\xi}_k^\top \mathbf{K}_k^{[k]} \underline{\xi}_k \middle| \mathcal{S}_k^{[k]}\right\} - \mathrm{E}\left\{\underline{\xi}_k^\top \mathbf{K}_k^{[k]} \underline{\xi}_k \middle| \mathcal{N}_k^{[k]}\right\} \\ &+ \sum_{i=k}^{k+N} \mathrm{E}\left\{\underline{\widehat{w}}_i^\top \mathbf{K}_i^{[k]} \underline{\widehat{w}}_i \middle| \mathcal{S}_k^{[k]}\right\} - \sum_{i=k}^{k+N} \mathrm{E}\left\{\underline{\widehat{w}}_i^\top \mathbf{K}_i^{[k]} \underline{\widehat{w}}_i \middle| \mathcal{N}_k^{[k]}\right\} + S_k \end{aligned}$$

PROOF. The proof is given in Appendix D. □

Finally, we briefly summarize the proposed event-triggered control law.

Event-Triggered Sequence-Based LQG Controller

The control law of the event-triggered sequence-based LQG controller consists of the following two steps that have to be evaluated at each time step k:

a) Calculate the control sequence candidate $\underline{U}_k^{cand}$ according to Theorem 5.1 .

b) Decide whether $\underline{U}_k^{cand}$ is sent or not sent to the actuator using the decision rule (5.16) by evaluating Lemma 5.1.

5.3. Performance and Stability Analysis

Similar to Section 3.5, we assess the stability of the closed-loop system based on the long run average costs. These are defined in analogy to Definition 3.3 for the sequence-based event-triggered LQG controller as

$$C_{\infty}^{Evt}(\underline{U}_0, \underline{U}_1, \ldots, s_0, s_1, \ldots) \stackrel{\text{def}}{=} \limsup_{K\to\infty} \frac{1}{K} \cdot C_{0\to K}^{Evt}\left(\underline{U}_{0:K-1}, s_{0:K-1}\right) , \tag{5.17}$$

where the expected cumulative costs $C_{0\to K}^{Evt}\left(\underline{U}_{0:K-1}\right)$ are defined in (5.1). In the style of Definition 3.4, the event-triggered LQG controller is said to be LRAC-stable if the long run average costs $C_{\infty}^{Evt}(\underline{U}_0, \underline{U}_1, \ldots, s_0, s_1, \ldots)$ are bounded. In the LRAC-stability analysis of the event-triggered system, we can use results already obtained for the S-LQG in Section 3.5. In particular, the long run average costs measured by the S-LQG cost function (3.3) can be used to constitute a lower and an upper bound for the long run average costs of the event-triggered controller.

We start with the derivation of the upper bound. By construction, the event-triggered controller does not send a sequence if, and only if, the expected cumulative costs are less than or equal to the expected cumulative costs incurred when the candidate sequence is sent at each time step. The optimal solution of the case in which a sequence is sent at each time step is the S-LQG. Now, if we measure the costs of the S-LQG in terms of the event-triggered cost function (5.17), it holds that

$$\begin{aligned} &C_{\infty}^{Evt}\left(\underline{U}_0^{S\text{-}LQG}, \underline{U}_1^{S\text{-}LQG}, \dots, s_0 = 1,\ s_1 = 1, \dots\right) \\ &= C_{\infty}^{S\text{-}LQG}\left(\underline{U}_0^{S\text{-}LQG}, \underline{U}_1^{S\text{-}LQG}, \dots\right) + S_k\ , \end{aligned} \tag{5.18}$$

where the upper index $S\text{-}LQG$ in $\underline{U}_k^{S\text{-}LQG}$ indicates that a control sequence is calculated using the S-LQG control law (see Theorem 3.1) and not with the event-triggered control law. The term $C_{\infty}^{S\text{-}LQG}(\cdots)$ accordingly refers to the long run average costs of the S-LQG given in Definition 3.3. The left hand side of (5.18) represents the costs if we use the S-LQG, but measure performance with the event-triggered cost function. Therefore, the decision variables are set to $s_{0:K} = 1$, as the S-LQG always sends sequences. Evaluating the left hand side and plugging $s_{0:K} = 1$ in (5.1), we see that the term

$$\limsup_{K \to \infty} \frac{1}{K} \cdot \sum_{k=0}^{K-1} (S_k s_k) = S_k \tag{5.19}$$

becomes constant and, hence, the equality (5.18) holds. The costs incurred by the event-triggered controller are therefore bounded from above by

$$C_{\infty}^{Evt}\left(\underline{U}_0^{Evt}, \underline{U}_1^{Evt}, \dots,\ s_0^{Evt},\ s_1^{Evt}, \dots\right) \leq C_{\infty}\left(\underline{U}_0^{SLQG}, \underline{U}_1^{SLQG}, \dots\right) + S_k\ ,$$

where we used the notation $\underline{U}_k^{Evt}$ and s_k^{Evt} to indicate that the corresponding values are calculated by the event-triggered control law.

Now, looking for a lower bound, we can make the observation that

$$C_{\infty}\left(\underline{U}_0^{SLQG}, \underline{U}_1^{SLQG}, \dots\right) \leq C_{\infty}^{Evt}\left(\underline{U}_0^{Evt}, \underline{U}_1^{Evt}, \dots\ ,\ s_0^{Evt},\ s_1^{Evt}, \dots\right)\ .$$

This inequality holds as the S-LQG is the optimal control law if a sequence is sent at each time step. Then, the S-LQG also obtains the minimum possible costs. Hence, neglecting the sending costs, i.e., dropping S_k at the right hand side of (5.18), yields a lower bound on the real minimum achievable costs.

As the long run average costs of the event-triggered case are lower and upper bounded by the long run average costs of the S-LQG (plus a finite term in the case of the lower bound), the stability conditions of the S-LQG derived in Section 3.5 are directly transferable to the event-based controller. We summarize this result in the following corollary.

Corollary 5.2 Consider the setup described in Section 5.1 with the event-triggered control law defined on page 97 . Then, the sufficient conditions in Theorem 3.3 and Theorem 3.4 for LRAC-stability and LRAC-instability of the S-LQG are also sufficient for the proposed event-triggered controller.

Unfortunately, the stability analysis does not reveal the stabilizing average data rate of the controller-actuator network that is required for the event-triggered controller to stabilize the system. It is obvious that the stabilizing average data rate of the event-triggered controller is smaller than the one needed for the S-LQG, but gaining more insights is still subject to research.

5.4. Evaluation

To evaluate the proposed approach, we use the inverted pendulum described in Section 4.3. The proposed event-triggered controller is compared to the controller described in [168]. The approach is chosen as it is one of the rare sequence-based controllers that can operate in an event-triggered mode while addressing time delays in the network connections. However, time delays are supposed to be bounded and no packet losses are considered. The sequence-based event-triggered approach described in [165] is not implemented here as the authors consider the sensor to be the decision maker, which is a different problem. The controller in [168] uses the sequence-based nominal controller method (see Section 2.4.2) to calculate a control sequence candidate $\underline{U}_k = [\underline{u}_{k|k}^\top, \underline{u}_{k+1|k}^\top, \ldots, \underline{u}_{k+N-1|k}^\top]^\top$. The candidate sequence is only sent to the actuator if it significantly differs from the previously sent sequence $\underline{U}_{k-\theta_k}$. The difference is determined by the normalized maximum norm according to

$$\max_{0\geq\ i\ \geq N-\theta_k} \frac{||\underline{u}_{k+i|k} - \underline{u}_{k+i|k-\theta_k}||_\infty}{||\underline{u}_{k+i|k}||_\infty} > \delta\ , \tag{5.20}$$

where $||\cdot||_\infty$ is the maximum norm and δ is a scalar design parameter, the so called *deadband.* The controller sends a sequence if either (5.20) is satisfied or the actuator is at risk of running out of applicable control inputs, i.e., if the value $N - \theta_k - 1$ is smaller than the maximum possible time delay in the controller-actuator network. Here, the nominal controller is implemented as an optimal LQG controller, and we use the two different values $\delta = 10$ and $\delta = 300$ for the deadband. The corresponding controllers are referred to as DB(10) and DB(300).

In the simulation, we consider two different networks with stochastic transmission characteristics as shown in Fig. 5.2. It can be seen that Network A provides a better transmission quality than Network B. For each network, the proposed event-triggered controller is designed as described on page 97, where we choose the sending costs $S_k = 1000$. In the following, this event-triggered S-LQG controller is abbreviated with ET-S-LQG. In addition, we also simulate the standard S-LQG controller that optimally compensates for network effects but sends a control sequence at each time step. We also implement an optimal LQG controller that is collocated at the pendulum. The LQG is not subject to network effects and also does not incur any sending costs. It serves as an ultimate lower bound on the costs to get a sense of the impact of network effects.

The described controllers have been simulated in 200 Monte Carlo simulation runs over 200 time steps for both networks and different sequence lengths. For each controller, each network, and each sequence length, three values are calculated: 1) the average transmission rate, 2) the average LQG costs without sending costs (3.16), and 3) the total costs (5.1) that consist of the LQG costs plus the sending costs. The sending costs are the average transmission rate multiplied with the weighting factor S_k.

The results are depicted in Fig. 5.3 and Fig. 5.4 for Network A and Network B, respectively. First, we investigate the results for Network A, which provides a good network quality. The total costs induced by the

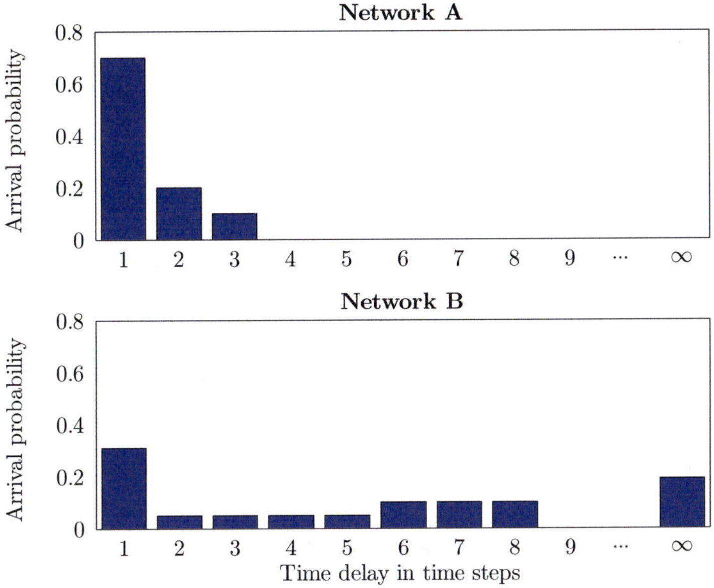

Figure 5.2.: Stochastic network characteristics used in the simulation

S-LQG are almost constant over the sequence lengths (upper plot). This is due to the good network quality; since time delays only have a small effect on performance, longer sequences do not yield a notable performance increase. Indeed, the difference between the S-LQG and the optimal LQG controller collocated at the plant is almost entirely caused by the network transmission costs that occur for the S-LQG, but not for the LQG. The performance of the S-LQG can therefore be significantly improved by the event-triggered approaches DB(10), DB(300), and ET-S-LQG that reduce the transmission costs component of the total costs. For these controllers, the costs reduce with increasing sequence length, because longer sequences can be used by the actuator for longer periods, which decreases the necessary transmission rate. Interestingly, for sequence lengths $N \geq 8$ all event-triggered approaches perform very similarly and even converge to almost the same costs at around 400.

The ET-S-LQG performs significantly better than DB(10) and DB(300) with small sequence lengths. This results from the facts that, first, the ET-S-LQG decides whether to send a sequence based on the expected total costs, rather than only on the change of the control inputs. Second, to ensure stability, the DB(10) and DB(300) have to send a control sequence as soon as the actuator is at risk of running out of applicable control inputs due to the underlying nominal-controller-based approach. Hence, the ET-S-LQG already reduces the communication rate for $N \geq 2$ (bottom plot), while DB(10) and DB(300) do not do this until $N \geq 4$. The early reduction of the ET-S-LQG raises the average costs in terms of the standard LQG cost function (middle plot), which measures the system performance ignoring the communication costs. However, this strategy is a good tradeoff in terms of the total costs (upper plot).

As seen in Fig. 5.4, the difference between the ET-S-LQG and the DB approaches becomes more distinct in Network B. Now, the ET-S-LQG shows a significantly better performance than DB(10) and DB(300) for short, as well as for long, sequence lengths. For small sequence lengths $N \leq 2$, the latter approaches are even unstable. This is a result of the bad network quality, since the approaches are not robust to packet losses. Considering long sequence lengths, we can see that DB(10) and DB(300) always send a sequence to the actuator, because the worse network characteristic triggers the sending condition at each time step. As a consequence, DB(10) and DB(300) perform almost identically. In the scenario with Network B, the ET-S-LQG strongly benefits from its cost-based decision rule that combines the properties of the S-LQG with the event-triggered control scheme. For small sequence lengths $N \leq 5$, the ET-S-LQG shows the same qualities as the S-LQG and is able to stabilize the plant despite enormous time delays and packet losses. For sequences with $N > 5$, the event-triggered component recognizes that communication can be reduced without risking stability. This allows the lowering of the transmission rate by a factor of 2.5. Therefore, the ET-S-LQG controller provides a reasonable tradeoff between stability and communication.

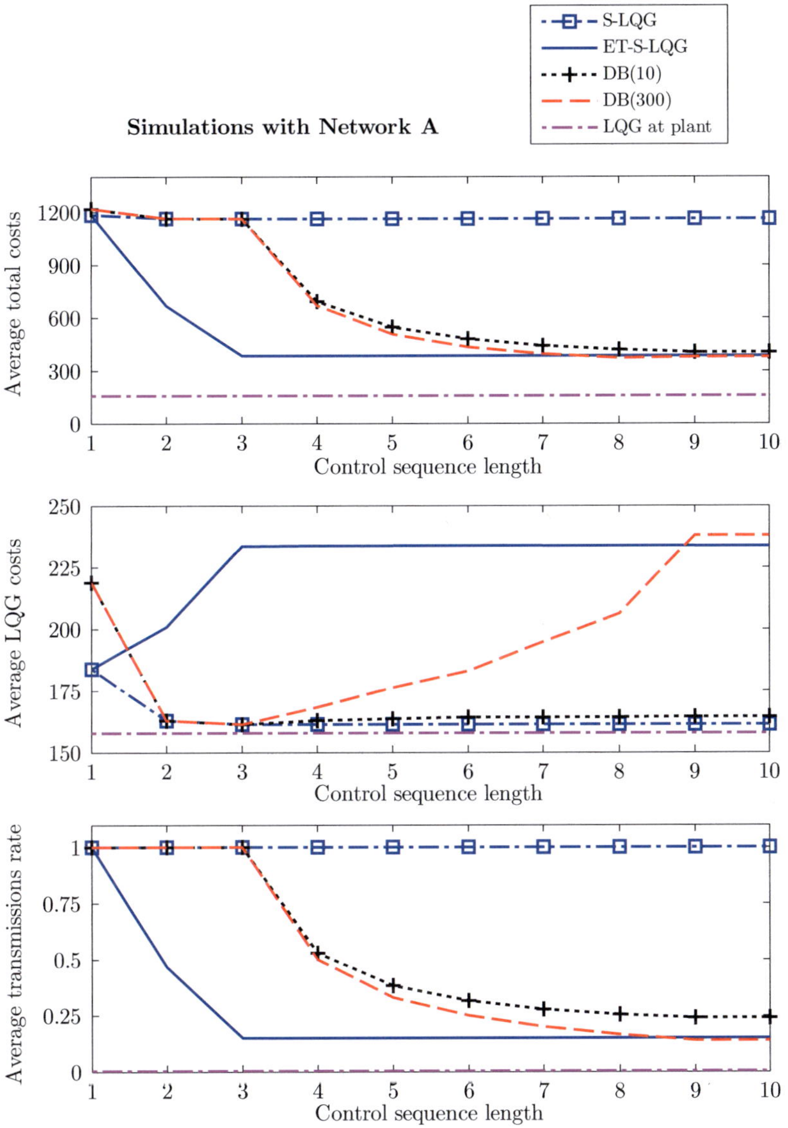

Figure 5.3.: Results of simulations with Network A

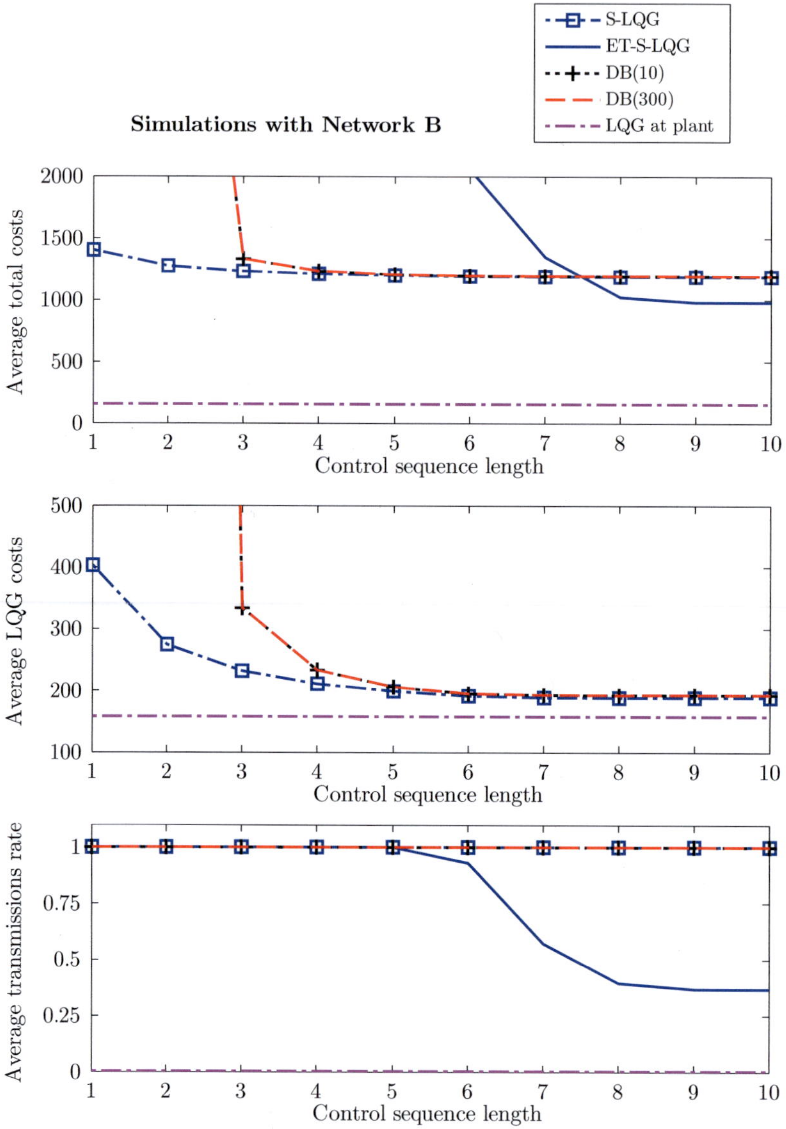

Figure 5.4.: Results of simulations with Network B

6. Conclusion

Since the dawn of the 21st century, considerable advances in wired and wireless communication networks have drastically enhanced the application range of feedback control systems on a global scale. However, the challenges to the use of this potential are tremendous. The networks can introduce time-varying transmission delays, packet losses, and band-limitations, which may elicit severe performance degradation and even render the feedback loop unstable. Therefore, the discipline of Networked Control Systems (NCS) emerged to address these problems in the intersection of communication and control. Among control methods discussed in the literature, sequence-based control promises an extremely high potential to cope with network-induced effects. The method exploits the packet-based transmission mode of modern communication networks by embedding additional information in the control data packets.

In this work, we approached the sequence-based method from the direction of stochastic optimal control and, in particular, from the foundation of LQG theory. We made the astonishing finding that the sequence-based counterpart of the LQG control problem can be optimally solved in analytic form despite the stochastic nonlinearity introduced by the communication networks. The optimal control law obtained is termed S-LQG, as it reveals much similarity to the conventional LQG control. In particular, we proved that the celebrated separation principle of LQG control preserves its validity in the sequenced-based NCS scenario as well. Benefits of the S-LQG include the linearity of the control with respect to the estimated state and the amenable convergence properties of the underlying feedback gains. As a result, the S-LQG requires relatively little computing power during operation and has a small memory demand. Further analysis of the S-LQG solution revealed new sufficient stability results for sequence-based NCS in regard to the boundedness of the long run average costs. Bearing in mind that the S-LQG is optimal, the criteria derived constitute fundamental bounds for the stabilizability of networked linear systems.

Widening the scope, we moved from the canonical stabilization problem to the higher level of tracking control and investigated how non-zero set points and entire reference trajectories can be brought together optimally with the sequence-based design philosophy. The result is the optimal sequence-based tracking controller presented in this work. This controller is not only able to compensate for time delays and packet losses in the network connections, but also, as a special feature, makes optimal use of available preview information on the future reference trajectory.

Finally, we focused our attention on the question of how the S-LQG controller can comply with additional band-limitations imposed by the networks. A possible answer has been found in the event-triggered extension of the S-LQG controller. In the event-triggered operation mode, the controller only occupies the network if necessary to ensure a prescribed performance level. The implemented strategy renders the event-triggered controller suboptimal, however, it allows for the preservation of the stability properties of the original S-LQG controller.

Future Research Directions

Despite the considerable effort that has been made to investigate the sequence-based stochastic optimal control problem, much remains to be done. One aspect is the robustness analysis of the S-LQG controller with respect to parameter uncertainties, unmodeled system dynamics, imperfect synchronization, etc. This will not only increase the practical applicability of the approach, but also offer a solid basis for robust sequence-based controller synthesis in general. Also, we only touched on the topic of sequence-based control when network acknowledgments are time delayed, or not available at all, as in UDP-like networks. Due to the immanently arising dual effect of the control, this problem class lacks the separation property of standard LQG control. It is widely recognized that this kind of stochastic optimization problem is very hard to solve. However, it is expected that future investigations will greatly benefit from current research activities in the field of approximate optimal control of Markov jump linear systems (MJLS) without mode observation. Coming from this direction, there is also great potential to gain more insights into the stochastic stability properties of the S-LQG and to find necessary stability criteria that complement the sufficient criteria derived in this work.

Moreover, nonlinear systems, distributed parameter systems, and decentralized control architectures are often encountered in NCS. Stochastic optimal control solutions for these problems rarely exist, even before applying the sequence-based method. Yet, the S-LQG can be a basis to synthesize sequence-based control solutions for these cases using approximate approaches such as the certainty equivalence design principle. Further, important extensions that could even permit an optimal solution within the S-LQG framework comprise the optimal compensation of quantization effects, consideration of H_∞-performance criteria, and the incorporation of integral linear constraints. In the literature, these problem classes have been proven to be compliant with the separation principle and, hence, are accessible to the stochastic optimal approach described in this work.

Finally, simulations demonstrate a very good performance of the S-LQG. However, it remains to be proven that this will also be the case in practice. Therefore, the practical application of the S-LQG has been initiated in conjunction with an industrial partner specializing in automation technology. The project aims for wireless motion control of a distributed electrical drive system in an industrial environment. Experimentation in this setup allows for the identification of bottlenecks of the S-LQG approach and quantification of the effects of network disturbances. It is expected the these practical results from this will trigger new theoretical research and, thus, will lead to an interactive feedback between theory and practice. Utilizing this huge feedback potential not only promises further practical improvements but also theoretical innovations and therefore should be a central direction of further research.

A. Proof of S-LQG Theorems

In this appendix, the proof of Theorem 3.1 is provided. In the first part of the proof, we integrate the MJLS description (3.29) into the optimization problem 3.1. This is done in Section A.1 by transformation of the cost function. Then, the optimal solution is derived in Section A.2 by application of the dynamic programming algorithm [13].

Remark A.1 In the literature, several solutions are available to the LQG control problem for MJLS [141]. However, these solutions cannot be applied directly in our case, because the mode θ_k is only available with a delay of one time step. Furthermore, measurements are subject to random transmission delays and packet losses. Finally, as we will see later, the weighting matrix for the control inputs of the augmented system is not positive definite, but becomes positive semidefinite.

In the derivation of the control law, we use the following lemma and assumption.

Lemma A.1 It holds that with any piecewise continuous function $g(\cdot)$ with at most countable number of discontinuities

$$\mathrm{E}\left\{\mathrm{E}\left\{g\left(\underline{\xi}_{k+1}\right)\middle|\mathcal{I}_{k+1}\right\}\middle|\mathcal{I}_k\right\} = \mathrm{E}\left\{g\left(\underline{\xi}_{k+1}\right)\middle|\mathcal{I}_k\right\} . \tag{A.1}$$

PROOF. The proof can be found in [67]. □

Assumption A.1 The control law is deterministically structured, i.e., if $\mathcal{I}_k$ contains deterministic values, then the controller output $\underline{U}_k = \mu_k(\mathcal{I}_k)$ is also deterministic (for all $k \leq K$).

A.1. Translation of Cost Function

We express the cost function (3.16) in terms of the augmented system state $\underline{\xi}_k$. With the definitions

$$\widehat{\mathbf{Q}}_K \stackrel{\text{def}}{=} \begin{bmatrix} \mathbf{Q}_K & \mathbf{0} \\ \mathbf{0} & \mathbf{0} \end{bmatrix} ,$$

$$\widehat{\mathbf{Q}}_k(\theta_k) \stackrel{\text{def}}{=} \begin{bmatrix} \mathbf{Q}_k & \mathbf{0} \\ \mathbf{0} & \mathbf{H}_k(\theta_k)^\top \mathbf{R}_k \mathbf{H}_k(\theta_k) \end{bmatrix} , \tag{A.2}$$

$$\widehat{\mathbf{R}}_k(\theta_k) \stackrel{\text{def}}{=} \mathbf{J}_k(\theta_k)^\top \mathbf{R}_k \mathbf{J}_k(\theta_k) , \tag{A.3}$$

it holds that

$$\underline{x}_K^\top \mathbf{Q}_K \underline{x}_K = \underline{\xi}_K^\top \widehat{\mathbf{Q}}_K \underline{\xi}_K ,$$

and

$$\begin{aligned}
&\underline{x}_k^\top \mathbf{Q}_k \underline{x}_k + \underline{u}_k^\top \mathbf{R}_k \underline{u}_k \\
&= \underline{x}_k^\top \mathbf{Q}_k \underline{x}_k + \left(\mathbf{H}_k(\theta_k)\underline{\rho}_k + \mathbf{J}_k(\theta_k)\underline{U}_k\right)^\top \mathbf{R}_k \left(\mathbf{H}_k(\theta_k)\underline{\rho}_k + \mathbf{J}_k(\theta_k)\underline{U}_k\right) \\
&\stackrel{(A.5)}{=} \underline{x}_k^\top \mathbf{Q}_k \underline{x}_k + \underline{\rho}_k^\top \mathbf{H}_k(\theta_k)^\top \mathbf{R}_k \mathbf{H}_k(\theta_k)\underline{\rho}_k + \underline{U}_k^\top \mathbf{J}_k(\theta_k)^\top \mathbf{R}_k \mathbf{J}_k(\theta_k)\underline{U}_k \\
&\quad + 2 \cdot \underline{U}_k^\top \mathbf{J}_k(\theta_k)^\top \mathbf{R}_k \mathbf{H}_k(\theta_k)\underline{\rho}_k \\
&\stackrel{(A.6)}{=} \underline{x}_k^\top \mathbf{Q}_k \underline{x}_k + \underline{\rho}_k^\top \mathbf{H}_k(\theta_k)^\top \mathbf{R}_k \mathbf{H}_k(\theta_k)\underline{\rho}_k + \underline{U}_k^\top \mathbf{J}_k(\theta_k)^\top \mathbf{R}_k \mathbf{J}_k(\theta_k)\underline{U}_k \\
&= \begin{pmatrix} \underline{x}_k \\ \underline{\rho}_k \end{pmatrix}^\top \begin{bmatrix} \mathbf{Q}_k & \mathbf{0} \\ \mathbf{0} & \mathbf{H}_k(\theta_k)^\top \mathbf{R}_k \mathbf{H}_k(\theta_k) \end{bmatrix} \begin{pmatrix} \underline{x}_k \\ \underline{\rho}_k \end{pmatrix} + \underline{U}_k^\top \mathbf{J}_k(\theta_k)^\top \mathbf{R}_k \mathbf{J}_k(\theta_k)\underline{U}_k \\
&= \underline{\xi}_k^\top \widehat{\mathbf{Q}}_k(\theta_k)\underline{\xi}_k + \underline{U}_k^\top \widehat{\mathbf{R}}_k(\theta_k)\underline{U}_k .
\end{aligned} \tag{A.4}$$

In the above derivation, we used the following lemma.

Lemma A.2 It holds that

$$\underline{U}_k^\top \mathbf{J}_k(\theta_k)^\top \mathbf{R}_k \mathbf{H}_k(\theta_k)\underline{\rho}_k = \underline{\rho}_k^\top \mathbf{H}_k(\theta_k)^\top \mathbf{R}_k \mathbf{J}_k(\theta_k)\underline{U}_k \tag{A.5}$$

$$= 0 . \tag{A.6}$$

PROOF. The equality (A.5) holds since the expression is scalar. The second equality follows due to the fact that for any $\theta_k \in \mathbb{J} = \{0, \dots, N\}$ it either holds that $\mathbf{J}_k(\theta_k) = \mathbf{0}$ or $\mathbf{H}_k(\theta_k) = \mathbf{0}$. □

Therefore, the cost function (3.16) can be written as

$$\begin{aligned} C_{0\to K} &= \underset{\substack{\underline{x}_{0:K} \\ \underline{u}_{0:K-1}}}{\mathrm{E}} \left\{ \underline{x}_K^\top \mathbf{Q}_K \underline{x}_K + \sum_{k=0}^{K-1} \underline{x}_k^\top \mathbf{Q}_k \underline{x}_k + \underline{u}_k^\top \mathbf{R}_k \underline{u}_k \middle| \mathcal{I}_0,\ \underline{U}_{0:K-1} \right\} \\ &= \underset{\substack{\underline{\xi}_{0:K} \\ \theta_{0:K-1}}}{\mathrm{E}} \left\{ \underline{\xi}_K^\top \widehat{\mathbf{Q}}_K \underline{\xi}_K + \sum_{k=0}^{K-1} \underline{\xi}_k^\top \widehat{\mathbf{Q}}_k(\theta_k) \underline{\xi}_k + \underline{U}_k^\top \widehat{\mathbf{R}}_k(\theta_k) \underline{U}_k \middle| \mathcal{I}_0,\ \underline{U}_{0:K-1} \right\} \end{aligned} \tag{A.7}$$

A.2. Minimization by Dynamic Programming

In the following, the transformed LQG costs (A.7) are minimized using dynamic programming. To that end, we introduce the minimum expected costs-to-go J_k^*. These describe the minimum expected cumulative costs attainable when operation would start at time step k and end at K.

Definition A.1 The minimum expected costs-to-go (from time step k to K) are defined by

$$J_k^* \overset{\text{def}}{=} \min_{\underline{U}_k} \underset{\substack{\underline{\xi}_k,\, \theta_k, \\ \mathcal{Y}_{k+1},\, \widehat{\underline{w}}_k}}{\mathrm{E}} \left\{ \underline{\xi}_k^\top \widehat{\mathbf{Q}}_k(\theta_k) \underline{\xi}_k + \underline{U}_k^\top \widehat{\mathbf{R}}_k(\theta_k) \underline{U}_k + J_{k+1}^* \middle| \mathcal{I}_k \right\} , \tag{A.8}$$

$$J_K^* \overset{\text{def}}{=} \underset{\underline{\xi}_K}{\mathrm{E}} \left\{ \underline{\xi}_K^\top \widehat{\mathbf{Q}}_K \underline{\xi}_K \middle| \mathcal{I}_K \right\} . \tag{A.9}$$

Bellman's principle of optimality states that an optimal solution to an optimization problem consists of optimal solutions of its subproblems [14]. In particular, it holds

$$J_0^* = C_{0\to K}^* , \tag{A.10}$$

where $C^*_{0\to K}$ are the minimum expected cumulative costs defined in (3.17). Hence, the optimal solution to the sequence-based LQG control problem can be obtained by recursively solving (A.8), starting with initial condition (A.9). In the following, we evaluate J^*_K, J^*_{K-1}, and J^*_{K-2}. It turns out that the subsequent expected cots-to-go J^*_{K-3} ... J^*_0 can then be solved via an inductive argument.

Time Step K

The minimal expected costs-to-go are directly given by (A.9). Introducing the definition $\mathbf{K}_K \stackrel{\text{def}}{=} \widehat{\mathbf{Q}}_K$, it holds that

$$J^*_K = \underset{\underline{\xi}_K}{\mathrm{E}} \left\{ \underline{\xi}_K^\top \widehat{\mathbf{Q}}_K \underline{\xi}_K \middle| \mathcal{I}_K \right\} = \underset{\underline{\xi}_K}{\mathrm{E}} \left\{ \underline{\xi}_K^\top \mathbf{K}_K \underline{\xi}_K \middle| \mathcal{I}_K \right\} .$$

Time Step $K-1$

According to (A.8), the minimal cost-to-go J^*_{K-1} are given by

$$\begin{aligned} J^*_{K-1} = \min_{\underline{U}_{K-1}} \mathrm{E} \Big\{ & \underline{\xi}_{K-1}^\top \widehat{\mathbf{Q}}_{K-1}(\theta_{K-1}) \underline{\xi}_{K-1} \\ & + \underline{U}_{K-1}^\top \widehat{\mathbf{R}}_{K-1}(\theta_{K-1}) \underline{U}_{K-1} + J^*_K \Big| \mathcal{I}_{K-1} \Big\} . \end{aligned}$$

Assumption A.1 allows us to explicitly condition on the minimization variable $\underline{U}_{K-1}$ so that

$$\begin{aligned} J^*_{K-1} &= \mathrm{E} \left\{ \underline{\xi}_{K-1}^\top \widehat{\mathbf{Q}}_{K-1}(\theta_{K-1}) \underline{\xi}_{K-1} \middle| \mathcal{I}_{K-1} \right\} && \text{(A.11)} \\ &\quad + \min_{\underline{U}_{K-1}} \left[\mathrm{E} \left\{ \underline{U}_{K-1}^\top \widehat{\mathbf{R}}_{K-1}(\theta_{K-1}) \underline{U}_{K-1} + \underline{\xi}_K^\top \mathbf{K}_K \underline{\xi}_K \middle| \mathcal{I}_{K-1}, \underline{U}_{K-1} \right\} \right] \\ &= \mathrm{E} \left\{ \underline{\xi}_{K-1}^\top \widehat{\mathbf{Q}}_{K-1}(\theta_{K-1}) \underline{\xi}_{K-1} \middle| \mathcal{I}_{K-1} \right\} && \text{(A.12)} \\ &\quad + \min_{\underline{U}_{K-1}} \Big[\underline{U}_{K-1}^\top \mathrm{E} \left\{ \widehat{\mathbf{R}}_{K-1}(\theta_{K-1}) \middle| \mathcal{I}_{K-1} \right\} \underline{U}_{K-1} \\ &\quad + \mathrm{E} \Big\{ \Big(\widehat{\mathbf{A}}_{K-1}(\theta_{K-1}) \underline{\xi}_{K-1} + \widehat{\mathbf{B}}_{K-1}(\theta_{K-1}) \underline{U}_{K-1} + \underline{\widehat{w}}_{K-1} \Big)^\top \mathbf{K}_K \end{aligned}$$

$$
\begin{aligned}
&\cdot \left(\widehat{\mathbf{A}}_{K-1}(\theta_{K-1})\underline{\xi}_{K-1} + \widehat{\mathbf{B}}_{K-1}(\theta_{K-1})\underline{U}_{K-1} + \underline{\widehat{w}}_{K-1}\right) \Big| \mathcal{I}_{K-1}, \underline{U}_{K-1} \Big\} \Big] \\
&= \mathrm{E}\left\{\underline{\xi}_{K-1}^{\top} \widehat{\mathbf{Q}}_{K-1}(\theta_{K-1})\underline{\xi}_{K-1} \Big| \mathcal{I}_{K-1}\right\} \\
&\quad + \min_{\underline{U}_{K-1}} \Big[\underline{U}_{K-1}^{\top} \mathrm{E}\left\{\widehat{\mathbf{R}}_{K-1}(\theta_{K-1}) \Big| \mathcal{I}_{K-1}\right\} \underline{U}_{K-1} \\
&\quad + \mathrm{E}\Big\{\underline{U}_{K-1}^{\top} \widehat{\mathbf{B}}_{K-1}(\theta_{K-1})^{\top} \mathbf{K}_K \widehat{\mathbf{B}}_{K-1}(\theta_{K-1}) \underline{U}_{K-1} \\
&\quad + 2 \cdot \underline{\xi}_{K-1}^{\top} \widehat{\mathbf{A}}_{K-1}(\theta_{K-1})^{\top} \mathbf{K}_K \widehat{\mathbf{B}}_{K-1}(\theta_{K-1}) \underline{U}_{K-1} \\
&\quad + \underline{\xi}_{K-1}^{\top} \widehat{\mathbf{A}}_{K-1}(\theta_{K-1})^{\top} \mathbf{K}_K \widehat{\mathbf{A}}_{K-1}(\theta_{K-1}) \underline{\xi}_{K-1} + \underline{\widehat{w}}_{K-1}^{\top} \mathbf{K}_K \underline{\widehat{w}}_{K-1} \\
&\quad + 2 \cdot \underline{\widehat{w}}_{K-1}^{\top} \mathbf{K}_K \left(\widehat{\mathbf{A}}_{K-1}(\theta_{K-1})\underline{\xi}_{K-1} + \widehat{\mathbf{B}}_{K-1}(\theta_{K-1})\underline{U}_{K-1}\right) \Big| \mathcal{I}_{K-1}, \underline{U}_{K-1} \Big\} \Big] \\
&= \mathrm{E}\left\{\underline{\xi}_{K-1}^{\top} \left(\widehat{\mathbf{Q}}_{K-1}(\theta_{K-1}) + \widehat{\mathbf{A}}_{K-1}(\theta_{K-1})^{\top} \mathbf{K}_K \widehat{\mathbf{A}}_{K-1}(\theta_{K-1})\right) \underline{\xi}_{K-1} \Big| \mathcal{I}_{K-1}\right\} \\
&\quad + \min_{\underline{U}_{K-1}} \Big[\underline{U}_{K-1}^{\top} \mathrm{E}\left\{\widehat{\mathbf{R}}_{K-1}(\theta_{K-1}) + \widehat{\mathbf{B}}_{K-1}(\theta_{K-1})^{\top} \mathbf{K}_K \widehat{\mathbf{B}}_{K}(\theta_{K}) \Big| \mathcal{I}_{K-1}\right\} \underline{U}_{K-1} \\
&\quad + 2 \cdot \mathrm{E}\left\{\underline{\xi}_{K-1}^{\top} \Big| \mathcal{I}_{K-1}\right\} \mathrm{E}\left\{\widehat{\mathbf{A}}_{K-1}(\theta_{K-1})^{\top} \mathbf{K}_K \widehat{\mathbf{B}}_{K-1}(\theta_{K-1}) \Big| \mathcal{I}_{K-1}\right\} \underline{U}_{K-1} \Big] \\
&\quad + \mathrm{E}\left\{\underline{\widehat{w}}_{K-1}^{\top} \mathbf{K}_K \underline{\widehat{w}}_{K-1} \Big| \mathcal{I}_{K-1}\right\} . \qquad \text{(A.13)}
\end{aligned}
$$

In the derivation above we used the fact that if $\mathcal{I}_{K-1}$ is given, the augmented state $\underline{\xi}_{K-1}$ is conditionally independent of θ_{K-1} and therefore of $\widehat{\mathbf{A}}_{K-1}(\theta_{K-1})$, $\widehat{\mathbf{B}}_{K-1}(\theta_{K-1})$, $\widehat{\mathbf{Q}}_{K-1}(\theta_{K-1})$, and $\widehat{\mathbf{R}}_{K-1}(\theta_{K-1})$.

As $\widehat{\mathbf{R}}_{K-1}(\theta_{K-1}) \geq 0$ and $\mathbf{K}_K \geq 0$, it can be seen from (A.12) that $J_{K-1}^* \geq 0$. Since J_{K-1}^* is also quadratic and convex in the control sequence $\underline{U}_{K-1}$, the minimum of J_{K-1}^* (with respect to $\underline{U}_{K-1}$) exists and can be computed by

$$
\begin{aligned}
\frac{\partial J_{K-1}^*}{\partial \underline{U}_{K-1}} &= 2 \cdot \mathrm{E}\left\{\widehat{\mathbf{R}}_{K-1}(\theta_{K-1}) + \widehat{\mathbf{B}}_{K-1}(\theta_{K-1})^{\top} \mathbf{K}_K \widehat{\mathbf{B}}_{K-1}(\theta_{K-1}) \Big| \mathcal{I}_{K-1}\right\} \underline{U}_{K-1} \\
&\quad + 2 \cdot \mathrm{E}\left\{\widehat{\mathbf{B}}_{K-1}(\theta_{K-1})^{\top} \mathbf{K}_K \widehat{\mathbf{A}}_{K-1}(\theta_{K-1}) \Big| \mathcal{I}_{K-1}\right\} \mathrm{E}\left\{\underline{\xi}_{K-1} \Big| \mathcal{I}_{K-1}\right\} \\
&\overset{!}{=} 0 . \qquad \text{(A.14)}
\end{aligned}
$$

The matrix $\mathrm{E}\left\{\widehat{\mathbf{R}}_{K-1}(\theta_{K-1}) + \widehat{\mathbf{B}}_{K-1}(\theta_{K-1})^\top \mathbf{K}_K \widehat{\mathbf{B}}_{K-1}(\theta_{K-1}) \middle| \mathcal{I}_{K-1}\right\}$ is not positive definite but only positive semidefinite. Therefore, the inverse does not exist and cannot be used to isolate $\underline{U}_{K-1}$. However, we can use the Moore-Penrose pseudoinverse instead, which is defined based on the *singular value decomposition* (SVD) [52]. For convenience, let us introduce the definition

$$\overline{\mathbf{R}}_{K-1} \stackrel{\text{def}}{=} \mathrm{E}\left\{\widehat{\mathbf{R}}_{K-1}(\theta_{K-1}) + \widehat{\mathbf{B}}_{K-1}(\theta_{K-1})^\top \mathbf{K}_K \widehat{\mathbf{B}}_{K-1}(\theta_{K-1}) \middle| \mathcal{I}_{K-1}\right\} . \tag{A.15}$$

The SVD of the positive semidefinite matrix $\overline{\mathbf{R}}_{K-1}$ is given by

$$\overline{\mathbf{R}}_{K-1} = \mathbf{S}_{K-1}^\top \begin{bmatrix} \mathbf{\Sigma}_{K-1} & \mathbf{0} \\ \mathbf{0} & \mathbf{0} \end{bmatrix} \mathbf{S}_{K-1} , \tag{A.16}$$

where $\mathbf{S}_{K-1}$ is a square orthogonal matrix, i.e., $\mathbf{S}_{K-1}^\top \mathbf{S}_{K-1} = \mathbf{I}$, and $\mathbf{\Sigma}_{K-1}$ is an invertible diagonal matrix that contains the positive singular values of $\overline{\mathbf{R}}_{K-1}$ [52]. Based on the SVD, the Moore-Penrose pseudoinverse $\overline{\mathbf{R}}_{K-1}^\dagger$ is given by

$$\overline{\mathbf{R}}_{K-1}^\dagger = \mathbf{S}_{K-1}^\top \begin{bmatrix} \mathbf{\Sigma}_{K-1}^{-1} & \mathbf{0} \\ \mathbf{0} & \mathbf{0} \end{bmatrix} \mathbf{S}_{K-1} . \tag{A.17}$$

Consequently, the condition (A.14) is equivalent to

$$\begin{aligned}
& \overline{\mathbf{R}}_{K-1} \underline{U}_{K-1} \\
& \quad = -\mathrm{E}\left\{\widehat{\mathbf{B}}_{K-1}(\theta_{K-1})^\top \mathbf{K}_K \widehat{\mathbf{A}}_{K-1}(\theta_{K-1}) \middle| \mathcal{I}_{K-1}\right\} \mathrm{E}\left\{\underline{\xi}_{K-1} \middle| \mathcal{I}_{K-1}\right\} \\
\Leftrightarrow \quad & \mathbf{S}_{K-1}^\top \begin{bmatrix} \mathbf{\Sigma}_{K-1}^{-1} & \mathbf{0} \\ \mathbf{0} & \mathbf{0} \end{bmatrix} \mathbf{S}_{K-1} \mathbf{S}_{K-1}^\top \begin{bmatrix} \mathbf{\Sigma}_{K-1} & \mathbf{0} \\ \mathbf{0} & \mathbf{0} \end{bmatrix} \mathbf{S}_{K-1} \underline{U}_{K-1} \\
& \quad = -\overline{\mathbf{R}}_{K-1}^\dagger \mathrm{E}\left\{\widehat{\mathbf{B}}_{K-1}(\theta_{K-1})^\top \mathbf{K}_K \widehat{\mathbf{A}}_{K-1}(\theta_{K-1}) \middle| \mathcal{I}_{K-1}\right\} \mathrm{E}\left\{\underline{\xi}_{K-1} \middle| \mathcal{I}_{K-1}\right\} \\
\Leftrightarrow \quad & \mathbf{S}_{K-1}^\top \begin{bmatrix} \mathbf{I} & \mathbf{0} \\ \mathbf{0} & \mathbf{0} \end{bmatrix} \mathbf{S}_{K-1} \underline{U}_{K-1} \\
& \quad = -\overline{\mathbf{R}}_{K-1}^\dagger \mathrm{E}\left\{\widehat{\mathbf{B}}_{K-1}(\theta_{K-1})^\top \mathbf{K}_K \widehat{\mathbf{A}}_{K-1}(\theta_{K-1}) \middle| \mathcal{I}_{K-1}\right\} \mathrm{E}\left\{\underline{\xi}_{K-1} \middle| \mathcal{I}_{K-1}\right\} .
\end{aligned}$$

This leads to the solution

$$\begin{aligned}
\underline{U}_{K-1} = {} & \overline{\mathbf{R}}_{K-1}^\dagger \cdot \mathrm{E}\left\{\widehat{\mathbf{B}}_{K-1}(\theta_{K-1})^\top \mathbf{K}_K \widehat{\mathbf{A}}_{K-1}(\theta_{K-1}) \middle| \mathcal{I}_{K-1}\right\} \mathrm{E}\left\{\underline{\xi}_{K-1} \middle| \mathcal{I}_{K-1}\right\} \\
& + \left(\mathbf{I} - \overline{\mathbf{R}}_{K-1}^\dagger \overline{\mathbf{R}}_{K-1}\right) \underline{a} ,
\end{aligned} \tag{A.18}$$

where $\underline{a}$ is an arbitrary vector. Without loss of generality, we can choose $\underline{a} = \underline{0}$ so that

$$\underline{U}_{K-1} = \overline{\mathbf{R}}_{K-1}^{\dagger} \cdot \mathrm{E}\left\{\widehat{\mathbf{B}}_{K-1}(\theta_{K-1})^{\top} \mathbf{K}_K \widehat{\mathbf{A}}_{K-1}(\theta_{K-1}) \middle| \mathcal{I}_{K-1}\right\} \mathrm{E}\left\{\underline{\xi}_{K-1} \middle| \mathcal{I}_{K-1}\right\} . \tag{A.19}$$

Remark A.2 Recently, a whole article has been dedicated to the problem of semidefinite weighting matrices and the associated discrete-time deterministic LQR control problem [31]. In the article, results and implications of using the Moore-Penrose pseudoinverse in the standard Riccati equation are discussed and a geometric interpretation is given.

Remark A.3 The equation (A.18) shows that some entries of the control sequence $\underline{U}_{K-1}$ are arbitrary, which technically results from the fact that $\overline{\mathbf{R}}_{K-1}$ is positive semidefinite and not positive definite as it would be in standard LQG control. The interpretation is that $\underline{U}_{K-1}$ contains control inputs that have no effect on the induced costs. That can have two reasons: First, if the network has a minimum latency that is longer than one time step, then the first control inputs of each sequence will never be applied. Second, the last N control sequences $\underline{U}_{K-1-N:K-1}$ contain control inputs (such as $\underline{u}_{K+1|K-1}$) that are supposed to be applied after the terminal time K. In this respect, the minimization problem is ill-posed as it contains minimization variables that have no influence on the cost function. One way to resolve this issue is to exclude the corresponding control inputs from the system equations, which, however, would lead to a time- and/or network dependent dimension of the system matrices. Another way is to use the Moore-Penrose pseudoinverse as shown above.

Using (A.19) in (A.13) restuls in

$$\begin{aligned}
J_{K-1}^{*} = {} & \mathrm{E}\left\{\underline{\xi}_{K-1}^{\top}\left(\widehat{\mathbf{Q}}_{K-1}(\theta_{K-1}) + \widehat{\mathbf{A}}_{K-1}(\theta_{K-1})^{\top} \mathbf{K}_K \widehat{\mathbf{A}}_{K-1}(\theta_{K-1})\right)\underline{\xi}_{K-1} \middle| \mathcal{I}_{K-1}\right\} \\
& - \mathrm{E}\left\{\underline{\xi}_{K-1}^{\top} \middle| \mathcal{I}_{K-1}\right\} \cdot \mathrm{E}\left\{\widehat{\mathbf{A}}_{K-1}(\theta_{K-1})^{\top} \mathbf{K}_K \widehat{\mathbf{B}}_{K-1}(\theta_{K-1}) \middle| \mathcal{I}_{K-1}\right\} \\
& \cdot \left(\mathrm{E}\left\{\widehat{\mathbf{R}}_{K-1}(\theta_{K-1}) + \widehat{\mathbf{B}}_{K-1}(\theta_{K-1})^{\top} \mathbf{K}_K \widehat{\mathbf{B}}_{K-1}(\theta_{K-1}) \middle| \mathcal{I}_{K-1}\right\}\right)^{\dagger} \\
& \cdot \mathrm{E}\left\{\widehat{\mathbf{B}}_{K-1}(\theta_{K-1})^{\top} \mathbf{K}_K \widehat{\mathbf{A}}_{K-1}(\theta_{K-1}) \middle| \mathcal{I}_{K-1}\right\} \mathrm{E}\left\{\underline{\xi}_{K-1} \middle| \mathcal{I}_{K-1}\right\}
\end{aligned}$$

$$+\,\mathrm{E}\left\{\underline{\widehat{w}}_{K-1}^{\top}\mathbf{K}_K\underline{\widehat{w}}_{K-1}\middle|\,\mathcal{I}_{K-1}\right\}\,.$$

We can write this in a more convenient form by defining

$$\begin{aligned}
\underline{e}_{K-1} &\stackrel{\text{def}}{=} \underline{\xi}_{K-1} - \mathrm{E}\left\{\underline{\xi}_{K-1}\middle|\,\mathcal{I}_{K-1}\right\}\,,\\
\mathbf{K}_{K-1} &\stackrel{\text{def}}{=} \mathrm{E}\left\{\widehat{\mathbf{Q}}_{K-1}(\theta_{K-1}) + \widehat{\mathbf{A}}_{K-1}(\theta_{K-1})^{\top}\mathbf{K}_K\widehat{\mathbf{A}}_{K-1}(\theta_{K-1})\middle|\,\mathcal{I}_{K-1}\right\} - \mathbf{P}_{K-1}\,,\\
\mathbf{P}_{K-1} &\stackrel{\text{def}}{=} \mathrm{E}\left\{\widehat{\mathbf{A}}_{K-1}(\theta_{K-1})^{\top}\mathbf{K}_K\widehat{\mathbf{B}}_{K-1}(\theta_{K-1})\middle|\,\mathcal{I}_{K-1}\right\} \qquad \text{(A.20)}\\
&\quad\cdot\left(\mathrm{E}\left\{\widehat{\mathbf{R}}_{K-1}(\theta_{K-1}) + \widehat{\mathbf{B}}_{K-1}(\theta_{K-1})^{\top}\mathbf{K}_K\widehat{\mathbf{B}}_{K-1}(\theta_{K-1})\middle|\,\mathcal{I}_{K-1}\right\}\right)^{\dagger}\\
&\quad\cdot\,\mathrm{E}\left\{\widehat{\mathbf{B}}_{K-1}(\theta_{K-1})^{\top}\mathbf{K}_K\widehat{\mathbf{A}}_{K-1}(\theta_{K-1})\middle|\,\mathcal{I}_{K-1}\right\}\,,
\end{aligned}$$

so that it yields

$$\begin{aligned}
J_{K-1}^{*} &= \mathrm{E}\left\{\underline{\widehat{w}}_{K-1}^{\top}\mathbf{K}_K\underline{\widehat{w}}_{K-1}\middle|\,\mathcal{I}_{K-1}\right\} + \mathrm{E}\left\{\underline{\xi}_{K-1}^{\top}\mathbf{K}_{K-1}\underline{\xi}_{K-1}\middle|\,\mathcal{I}_{K-1}\right\}\\
&\quad+\mathrm{E}\left\{\underline{e}_{K-1}^{\top}\mathbf{P}_{K-1}\underline{e}_{K-1}\middle|\,\mathcal{I}_{K-1}\right\}\,.
\end{aligned}$$

Time Step $K-2$

According to (A.8) and using Assumption A.1, the minimal expected costs-to-go at time step $K-2$ are given by

$$\begin{aligned}
J_{K-2}^{*} &= \min_{\underline{U}_{K-2}}\Bigg[\mathrm{E}\Big\{\underline{\xi}_{K-2}^{\top}\widehat{\mathbf{Q}}_{K-2}(\theta_{K-2})\underline{\xi}_{K-2}\\
&\quad+\underline{U}_{K-2}^{\top}\widehat{\mathbf{R}}_{K-2}(\theta_{K-2})\underline{U}_{K-2} + J_{K-1}^{*}\Big|\,\mathcal{I}_{K-2},\underline{U}_{K-2}\Big\}\Bigg]\\
&= \mathrm{E}\left\{\underline{\xi}_{K-2}^{\top}\widehat{\mathbf{Q}}_{K-2}(\theta_{K-2})\underline{\xi}_{K-2}\middle|\,\mathcal{I}_{K-2}\right\} \qquad \text{(A.21)}\\
&\quad+\min_{\underline{U}_{K-2}}\Bigg[\underline{U}_{K-2}^{\top}\mathrm{E}\left\{\widehat{\mathbf{R}}_{K-2}(\theta_{K-2})\middle|\,\mathcal{I}_{K-2}\right\}\underline{U}_{K-2}\\
&\quad+\mathrm{E}\left\{\underline{\xi}_{K-1}^{\top}\mathbf{K}_{K-1}\underline{\xi}_{K-1}\middle|\,\mathcal{I}_{K-2},\underline{U}_{K-2}\right\}\Bigg]\\
&\quad+\mathrm{E}\left\{\underline{e}_{K-1}^{\top}\mathbf{P}_{K-1}\underline{e}_{K-1}\middle|\,\mathcal{I}_{K-2}\right\} + \mathrm{E}\left\{\underline{\widehat{w}}_{K-1}^{\top}\mathbf{K}_K\underline{\widehat{w}}_{K-1}\middle|\,\mathcal{I}_{K-2}\right\}.
\end{aligned}$$

The term $\mathrm{E}\left\{\underline{e}_{K-1}^{\top}\mathbf{P}_{K-1}\underline{e}_{K-1}\middle|\,\mathcal{I}_{K-2}\right\}$ penalizes the weighted estimation error covariance expected in the next time step. This is reasonable as a higher uncertainty about the future state will very likely impose additional costs. A key step in the derivation of the optimal control law is to see that $\mathrm{E}\left\{\underline{e}_{K-1}^{\top}\mathbf{P}_{K-1}\underline{e}_{K-1}\middle|\,\mathcal{I}_{K-2}\right\}$ is independent of the control sequence $\underline{U}_{K-2}$ and therefore can be excluded from the minimization (as already done in (A.21)). The proof of this proposition is given in the following Lemma A.3.

Lemma A.3 The weighted one step prediction of the estimation error covariance $\mathrm{E}\left\{\underline{e}_{k}^{\top}\mathbf{P}_{k}\underline{e}_{k}\middle|\,\mathcal{I}_{k-1},\underline{U}_{k-1}\right\}$ is independent of the control sequences $\underline{U}_{0:k-1}$ that are sent. In other words,

$$\mathrm{E}\left\{\underline{e}_{k}^{\top}\mathbf{P}_{k}\underline{e}_{k}\middle|\,\mathcal{I}_{k-1},\underline{U}_{k-1}\right\} = \mathrm{E}\left\{\underline{e}_{k}^{\top}\mathbf{P}_{k}\underline{e}_{k}\middle|\,\{\mathcal{I}_{k-1}\backslash\underline{U}_{0:k}\}\right\} .$$

Proof. The proof follows the idea of Lemma 5.2.1 in [14] where the independence of the estimation error was proven for the standard LQG control problem. In the following, we show that the induced estimation error of the controlled system at hand is identical to the estimation error of the associated uncontrolled system. This directly implies that the estimation error must be independent of the control input sequence.

Consider the system (3.29)

$$\underline{\xi}_{k+1} = \widehat{\mathbf{A}}_k(\theta_k)\,\underline{\xi}_k + \widehat{\mathbf{B}}_k(\theta_k)\,\underline{U}_k + \underline{\widehat{w}}_k \ ,$$

and the associated uncontrolled system with state $\underline{\xi}_k^{un}$

$$\underline{\xi}_{k+1}^{un} = \widehat{\mathbf{A}}_k(\theta_k)\,\underline{\xi}_k^{un} + \underline{\widehat{w}}_k \ , \tag{A.22}$$

that has the same system matrices, initial conditions, noise realizations $\underline{\widehat{w}}_{0:k-1}$ and network delay realizations $\tau_{0:k-1}^{CA}$, $\tau_{0:k-1}^{SC}$. Both systems evolve according to linear transformations. It is therefore possible to find deterministic matrices $\mathbf{\Omega}_k$, $\mathbf{\Psi}_k$, and $\mathbf{\Lambda}_k$ depending on $\theta_{0:k-1}$ such that

$$\begin{aligned}
\underline{\xi}_k &= \mathbf{\Omega}_k\underline{\xi}_0 + \mathbf{\Psi}_k\left[\underline{U}_0^{\top},\ldots,\underline{U}_{k-1}^{\top}\right]^{\top} + \mathbf{\Lambda}_k\left[\underline{\widehat{w}}_0^{\top},\cdots,\underline{\widehat{w}}_{k-1}^{\top}\right]^{\top} \ ,\\
\underline{\xi}_k^{un} &= \mathbf{\Omega}_k\underline{\xi}_0 + \mathbf{\Lambda}_k\left[\underline{\widehat{w}}_0^{\top},\cdots,\underline{\widehat{w}}_{k-1}^{\top}\right]^{\top} .
\end{aligned}$$

Then, we obtain for the expected values

$$\mathrm{E}\left\{\underline{\xi}_k \middle| \mathcal{I}_k\right\} = \mathbf{\Omega}_k \cdot \mathrm{E}\left\{\underline{\xi}_0 \middle| \mathcal{I}_k\right\} + \mathbf{\Psi}_k \left[\underline{U}_0^\top, \cdots, \underline{U}_{k-1}^\top\right]^\top ,$$
$$\mathrm{E}\left\{\underline{\xi}_k^{un} \middle| \mathcal{I}_k\right\} = \mathbf{\Omega}_k \cdot \mathrm{E}\left\{\underline{\xi}_0 \middle| \mathcal{I}_k\right\} ,$$

where $\mathbf{\Omega}_k$ and $\mathbf{\Psi}_k$ are known since information vector $\mathcal{I}_k$ includes $\theta_{0:k-1}$. The corresponding estimation errors

$$\underline{e}_k \overset{\text{def}}{=} \underline{\xi}_k - \mathrm{E}\left\{\underline{\xi}_k \middle| \mathcal{I}_k\right\} ,$$
$$\underline{e}_k^{un} \overset{\text{def}}{=} \underline{\xi}_k^{un} - \mathrm{E}\left\{\underline{\xi}_k^{un} \middle| \mathcal{I}_k\right\} ,$$

are therefore given by

$$\underline{e}_k = \mathbf{\Omega}_k \left(\underline{\xi}_0 - \mathrm{E}\left\{\underline{\xi}_0 \middle| \mathcal{I}_k\right\}\right) + \mathbf{\Lambda}_k \left(\underline{\widehat{w}}_0^\top \quad \cdots \quad \underline{\widehat{w}}_{k-1}^\top\right)^\top ,$$
$$\underline{e}_k^{un} = \mathbf{\Omega}_k \left(\underline{\xi}_0 - \mathrm{E}\left\{\underline{\xi}_0 \middle| \mathcal{I}_k\right\}\right) + \mathbf{\Lambda}_k \left(\underline{\widehat{w}}_0^\top \quad \cdots \quad \underline{\widehat{w}}_{k-1}^\top\right)^\top .$$

We see that the estimation errors are identical and, consequently, $\underline{e}_k$ must be independent of $\underline{U}_{0:k-1}$. Noting that according to (A.20) the expected weighting matrix

$$\mathrm{E}\left\{\mathbf{P}_k \middle| \mathcal{I}_{k-1}, \underline{U}_{k-1}\right\} = \mathrm{E}\left\{\mathbf{P}_k \middle| \theta_{k-2}\right\}$$

is also independent of $\underline{U}_{0:k-1}$, concludes the proof. □

Lemma A.3 proves that the expected estimation error and its covariance are independent of the control sequences. This implies that the separation theorem known from standard LQG control also extends to the sequence-based setup with TCP-like networks. The separation theorem states that the optimal controller can be separated without loss of optimality into an optimal state estimator and into an optimal state-feedback controller (see also the discussion in Section 3.4). In [55, 120], it is shown that separation holds for packet-dropping TCP-like networks with optimal non-sequence-based controllers. Lemma A.3 extends these results to the sequence-based setup and the presence of time-varying transmission delays. It is worth pointing out that the Assumption 3.5 of a TCP-like connection is crucial for Lemma A.3 to hold.

Comparing (A.11) and (A.21), it can be seen that the structure of these equations is the same with the exception of two additive terms that are independent of $\underline{U}_{0:k-1}$. Hence, minimizing over $\underline{U}_{K-2}$ will result in costs-to-go J^*_{K-2} of the same structure as J^*_{K-1}. Therefore, it follows that by an inductive argument for control sequence $\underline{U}_k$ and cost-to-go J^*_k

$$\underline{U}_k = \mathbf{L}_k \cdot \mathrm{E}\left\{\underline{\xi}_k \middle| \mathcal{I}_k\right\} , \tag{A.23}$$

$$\begin{aligned} J^*_k &= \sum_{i=k}^{K-1} \mathrm{E}\left\{\widehat{\underline{w}}_i^\top \mathbf{K}_{i+1} \widehat{\underline{w}}_i \middle| \mathcal{I}_k\right\} + \mathrm{E}\left\{\underline{\xi}_k^\top \mathbf{K}_{k+1} \underline{\xi}_k \middle| \mathcal{I}_k\right\} \\ &+ \sum_{i=k}^{K-1} \mathrm{E}\left\{\underline{e}_i^\top \mathbf{P}_i \underline{e}_i \middle| \mathcal{I}_k\right\} , \end{aligned} \tag{A.24}$$

with

$$\begin{aligned} \mathbf{L}_k = &\left(\mathrm{E}\left\{\widehat{\mathbf{R}}_k(\theta_k) + \widehat{\mathbf{B}}_k(\theta_k)^\top \mathbf{K}_{k+1} \widehat{\mathbf{B}}_k(\theta_k) \middle| \mathcal{I}_k\right\}\right)^\dagger \\ &\cdot \mathrm{E}\left\{\widehat{\mathbf{B}}_k(\theta_k)^\top \mathbf{K}_{k+1} \widehat{\mathbf{A}}_k(\theta_k) \middle| \mathcal{I}_k\right\} , \end{aligned} \tag{A.25}$$

$$\begin{aligned} \mathbf{K}_k = \mathrm{E}&\left\{\widehat{\mathbf{Q}}_k(\theta_k) + \widehat{\mathbf{A}}_k(\theta_k)^\top \mathbf{K}_{k+1} \widehat{\mathbf{A}}_k(\theta_k) \middle| \mathcal{I}_k\right\} \\ &- \mathrm{E}\left\{\widehat{\mathbf{A}}_k(\theta_k)^\top \mathbf{K}_{k+1} \widehat{\mathbf{B}}_k(\theta_k) \middle| \mathcal{I}_k\right\} \\ &\cdot \left(\mathrm{E}\left\{\widehat{\mathbf{R}}_k(\theta_k) + \widehat{\mathbf{B}}_k(\theta_k)^\top \mathbf{K}_{k+1} \widehat{\mathbf{B}}_k(\theta_k) \middle| \mathcal{I}_k\right\}\right)^\dagger \\ &\cdot \mathrm{E}\left\{\widehat{\mathbf{B}}_k(\theta_k)^\top \mathbf{K}_{k+1} \widehat{\mathbf{A}}_k(\theta_k) \middle| \mathcal{I}_k\right\} . \end{aligned} \tag{A.26}$$

The conditional expectations in (A.25) and (A.26) can be computed because $\mathbf{L}_k$ and $\mathbf{K}_k$ only depend on θ_{k-1}, which is part of the information set $\mathcal{I}_k$, and not on $\theta_{0:k-2}$ or $\theta_{k-1:K}$. To emphasize this dependency, we write $\mathbf{L}_k = \mathbf{L}_k(\theta_{k-1})$ and $\mathbf{K}_k = \mathbf{K}_k(\theta_{k-1})$. Now, we can use the law of total probability by explicitly conditioning on a specific value of θ_{k-1} such as $\theta_{k-1} = i$ with $i \in \mathbb{J}$. This way, we obtain

$$\begin{aligned}\mathbf{L}_k(i) = &-\left(\sum_{j=0}^{N} p(i,j)\left[\widehat{\mathbf{R}}_k(j) + \widehat{\mathbf{B}}_k(j)^\top \mathbf{K}_{k+1}(j)\widehat{\mathbf{B}}_k(j)\right]\right)^\dagger \\ &\cdot \left(\sum_{j=0}^{N} p(i,j)\left[\widehat{\mathbf{B}}_k(j)^\top \mathbf{K}_{k+1}(j)\widehat{\mathbf{A}}_k(j)\right]\right)\end{aligned} \tag{A.27}$$

and

$$\begin{aligned}\mathbf{K}_k(i) = &\sum_{j=0}^{N} p(i,j)\left[\widehat{\mathbf{Q}}_k(j) + \widehat{\mathbf{A}}_k(j)^\top \mathbf{K}_{k+1}(j)\widehat{\mathbf{A}}_k(j)\right] \\ &- \left(\sum_{j=0}^{N} p(i,j)\left[\widehat{\mathbf{A}}_k(j)^\top \mathbf{K}_{k+1}(j)\widehat{\mathbf{B}}_k(j)\right]\right) \\ &\cdot \left(\sum_{j=0}^{N} p(i,j)\left[\widehat{\mathbf{R}}_k(j) + \widehat{\mathbf{B}}_k(j)^\top \mathbf{K}_{k+1}(j)\widehat{\mathbf{B}}_k(j)\right]\right)^\dagger \\ &\cdot \left(\sum_{j=0}^{N} p(i,j)\left[\widehat{\mathbf{B}}_k(j)^\top \mathbf{K}_{k+1}(j)\widehat{\mathbf{A}}_k(j)\right]\right),\end{aligned} \tag{A.28}$$

with $p(i,j)$ denoting the elements of transition matrix $\mathbf{T}$ defined in Lemma 3.1. This concludes the proof of Theorem 3.1 .

B. Proof of Stability Theorems

This section provides proofs for the stability Theorems 3.3 , 3.4 , and Lemma 3.2. The conditions in Theorem 3.3 a) and Theorem 3.4 a), which refer to the boundedness and unboundedness of the error covariance matrix, can be found in [118] and therefore are not repeated here. The other proofs are based on [45, 126], where the boundedness of the long run average costs is only investigated for the case of packet losses.

In the derivation, we express the recursion (3.32) via the operator $g(\cdot)$ such that

$$\mathbf{K}_k(0:N) = g(\mathbf{K}_{k+1}(0:N)) \ , \tag{B.1}$$

where

$$g(\mathbf{X}(0:N)) \stackrel{\text{def}}{=} (g_0(\mathbf{X}(0:N)), \ldots, g_N(\mathbf{X}(0:N))) \ , \tag{B.2}$$

and

$$\begin{aligned}
&g_j(\mathbf{X}(0:N)) \stackrel{\text{def}}{=} \\
&\left[\sum_{i=0}^{N} p(j,i)\left(\widehat{\mathbf{Q}}(i) + \widehat{\mathbf{A}}(i)^\top \mathbf{X}(i)\widehat{\mathbf{A}}(i)\right)\right] - \left[\sum_{i=0}^{N} p(j,i)\widehat{\mathbf{A}}(i)^\top \mathbf{X}(i)\widehat{\mathbf{B}}(i)\right] \\
&\times \left[\sum_{i=0}^{N} p(j,i)\left(\widehat{\mathbf{R}}(i) + \widehat{\mathbf{B}}(i)^\top \mathbf{X}(i)\widehat{\mathbf{B}}(i)\right)\right]^\dagger \left[\sum_{i=0}^{N} p(j,i)\widehat{\mathbf{B}}(i)^\top \mathbf{X}(i)\widehat{\mathbf{A}}(i)\right] .
\end{aligned}$$

In (B.2), $g : \mathbb{R}^{n_\xi \times n_\xi}, \ldots, \mathbb{R}^{n_\xi \times n_\xi} \rightarrow \mathbb{R}^{n_\xi \times n_\xi}, \ldots, \mathbb{R}^{n_\xi \times n_\xi}$ is an operator that maps a sequence of $N+1$ matrices $\mathbf{X}(0:N) = (\mathbf{X}(0), \ldots, \mathbf{X}(N))$ to a sequence of $N+1$ matrices with the same dimension. Further, the operator $g_j : \mathbb{R}^{n_\xi \times n_\xi}, \ldots, \mathbb{R}^{n_\xi \times n_\xi} \rightarrow \mathbb{R}^{n_\xi \times n_\xi}$ maps a sequence of $N+1$ square matrices to a matrix with the same dimension as $\mathbf{X}(j)$.

Based on this description, the following lemma formulates the important observation that the control related costs can be treated separately from the estimation error related costs when analyzing LRAC-stability.

Lemma B.1 Consider the NCS setup described in Section 3.1 with Assumption 3.6 controlled by the S-LQG given in Theorem 3.1 . Then, the system is LRAC-stable for all initial conditions $(\underline{\overline{x}}_0, \mathbf{\Lambda}_0)$ if and only if

a) the sequence $\mathbf{X}_{k+1}(0:N) = g(\mathbf{X}_k(0:N))$ is upper bounded, and

b) the expected estimation error covariance $\mathrm{E}\left\{\underline{e}_k \underline{e}_k^\top \,\middle|\, \mathcal{I}_0\right\}$ is upper bounded.

Proof. Note that for any symmetric random matrix $\mathbf{Z}$ and any random vector $\underline{z}$ that are stochastically independent of each other, it holds that

$$\mathrm{E}\left\{\underline{z}^\top \mathbf{Z} \underline{z}\right\} = \mathrm{tr}\left(\mathrm{E}\left\{\mathbf{Z}\right\} \mathrm{E}\left\{\underline{z}\underline{z}^\top\right\}\right) + \mathrm{E}\left\{\underline{z}^\top\right\} \mathrm{E}\left\{\mathbf{Z}\right\} \mathrm{E}\left\{\underline{z}\right\} . \tag{B.3}$$

Using (B.3), (3.33), and Definition 3.3, it holds that for the long run average costs $C_\infty^{S\text{-}LQG}$ induced by the S-LQG

$$\begin{aligned} C_\infty^{S\text{-}LQG} &= \limsup_{K\to\infty} \frac{1}{K} \cdot C^*_{0\to K} \\ &= \limsup_{K\to\infty} \frac{1}{K} \cdot \Bigg[\mathrm{tr}\left(\mathrm{E}\left\{\mathbf{K}_0(N) | \mathcal{I}_0\right\} \begin{bmatrix} \mathbf{\Lambda}_0 & \underline{0} \\ \underline{0} & \underline{0} \end{bmatrix} \right) + \begin{bmatrix} \underline{\overline{x}}_0 \\ \underline{0} \end{bmatrix}^\top \mathrm{E}\left\{\mathbf{K}_0\right\} \begin{bmatrix} \underline{\overline{x}}_0 \\ \underline{0} \end{bmatrix} \\ &+ \sum_{k=0}^{K-1} \mathrm{tr}\left(\mathrm{E}\left\{ \widehat{\mathbf{Q}}_k(\theta_k) + \widehat{\mathbf{A}}_k(\theta_k)^\top \mathbf{K}_{k+1}(\theta_k) \widehat{\mathbf{A}}_k(\theta_k) \right.\right. \\ &\left.\left. - \mathbf{K}_k(\theta_{k-1}) | \mathcal{I}_0 \right\} \mathrm{E}\left\{\underline{e}_k \underline{e}_k^\top \,\middle|\, \mathcal{I}_0\right\} \right) + \sum_{k=0}^{K-1} \mathrm{tr}\left(\mathrm{E}\left\{\mathbf{K}_{k+1}(\theta_k) | \mathcal{I}_0\right\} \mathbf{W}\right) \Bigg] . \end{aligned}$$

As the matrices $\widehat{\mathbf{Q}}(\theta_k)$, $\widehat{\mathbf{A}}(\theta_k)$, $\mathbf{\Lambda}_0$, and $\mathbf{W}$ are bounded, the only possibilities for unboundedness of (B.4) are that the matrices $\mathrm{E}\{\mathbf{K}_k(0:N)|\mathcal{I}_k\}$ or $\mathrm{E}\left\{\underline{e}_k \underline{e}_k^\top \,\middle|\, \mathcal{I}_0\right\}$ grow unbounded. With (B.1), this proves the necessity of Lemma B.1 a) and b). Sufficiency of Lemma B.1 a) can be shown by assuming that (B.1) is unbounded. Then, $\mathbf{X}_{k+1}(0:N) = g(\mathbf{X}_k(0:N))$

grows unbounded and together with Assumption 3.6 d) that $(\mathbf{A}, \mathbf{W}^{1/2})$ is controllable, it follows that

$$\limsup_{K\to\infty} \frac{1}{K} \cdot \sum_{k=0}^{K-1} \operatorname{tr}\left(\mathrm{E}\left\{\mathbf{K}_{k+1}(\theta_k)\middle| \mathcal{I}_0\right\} \mathbf{W}\right) \tag{B.4}$$

grows unbounded. To prove sufficiency of Lemma B.1 b), we use the fact that

$$\begin{aligned}
\mathrm{E}\left\{\underline{e}_k\underline{e}_k^\top \middle| \mathcal{I}_0\right\} &= \mathrm{E}\left\{\left(\underline{\xi}_k - \mathrm{E}\left\{\underline{\xi}_k\middle|\mathcal{I}_0\right\}\right)\left(\underline{\xi}_k - \mathrm{E}\left\{\underline{\xi}_k\middle|\mathcal{I}_0\right\}\right)^\top\middle|\mathcal{I}_0\right\} \\
&= \mathrm{E}\left\{\underline{\xi}_k\underline{\xi}_k^\top\middle|\mathcal{I}_0\right\} - \mathrm{E}\left\{\underline{\xi}_k\middle|\mathcal{I}_0\right\}\mathrm{E}\left\{\underline{\xi}_k^\top\middle|\mathcal{I}_0\right\} .
\end{aligned} \tag{B.5}$$

All terms of (B.5) are positive semidefinite. Hence, $\mathrm{E}\{\underline{\xi}_k\underline{\xi}_k^\top|\mathcal{I}_0\}$ must grow unbounded for $k \to \infty$ if $\mathrm{E}\left\{\underline{e}_k\underline{e}_k^\top\middle|\mathcal{I}_0\right\}$ is unbounded. Using (3.3) and (3.16), this implies that the costs must be unbounded due to Assumption 3.6 c) which concludes the proof. □

We now introduce three additional operators that will be used to constitute an upper and a lower bound for the recursion (B.2).

$$\begin{aligned}
\Phi_j(\widehat{\mathbf{L}}(j), \mathbf{X}(0:N)) &\overset{\text{def}}{=} \sum_{i=0}^{N} p(j,i)\widehat{\mathbf{Q}}(i) + \sum_{i=0}^{N} p(j,i)\widehat{\mathbf{L}}(j)\widehat{\mathbf{R}}(i)\widehat{\mathbf{L}}(j) \\
&+ \sum_{i=0}^{N} p(j,i)\left(\widehat{\mathbf{A}}(i) + \widehat{\mathbf{B}}(i)\widehat{\mathbf{L}}(j)\right)^\top \mathbf{X}(i)\left(\widehat{\mathbf{A}}(i) + \widehat{\mathbf{B}}(i)\widehat{\mathbf{L}}(j)\right) ,
\end{aligned} \tag{B.6}$$

$$\begin{aligned}
L_j^\Phi(\mathbf{X}(0:N)) \overset{\text{def}}{=} &- \left[\sum_{i=0}^{N} p(j,i)\left(\widehat{\mathbf{R}}(i) + \widehat{\mathbf{B}}(i)^\top\mathbf{X}(i)\widehat{\mathbf{B}}(i)\right)\right]^\dagger \\
&\times \left[\sum_{i=0}^{N} p(j,i)\widehat{\mathbf{B}}(i)^\top\mathbf{X}(i)\widehat{\mathbf{A}}(i)\right] ,
\end{aligned} \tag{B.7}$$

$$\Gamma(\mathbf{X}) \overset{\text{def}}{=} \Phi_N\left(L_N^\Phi(\mathbf{0},\ldots,\mathbf{0},\mathbf{X}), (\mathbf{0},\ldots,\mathbf{0},\mathbf{X})\right) . \tag{B.8}$$

The following Lemmas summarize useful properties of these operators.

Lemma B.2 Assume that $\mathbf{X}(0:N) \geq 0$, then it holds:

a) $\arg\min_{\widehat{\mathbf{L}}(j)} \Phi_j(\widehat{\mathbf{L}}(j), \mathbf{X}(0:N)) = L_j^{\Phi}(\mathbf{X}(0:N))$,

b) $\min_{\widehat{\mathbf{L}}(j)} \Phi_j(\widehat{\mathbf{L}}(j), \mathbf{X}(0:N)) = \Phi_j(L_j^{\Phi}(\mathbf{X}(0:N)), \mathbf{X}(0:N))$
$= g_j(\mathbf{X}(0:N))$,

c) $g_j(\mathbf{X}(0:N)) \leq \Phi_j(\mathbf{L}(j), \mathbf{X}(0:N)) \quad, \forall\, \mathbf{L}(j)$,

d) if $\mathbf{X}(0:N) \geq \mathbf{Y}(0:N)$, then $g_j(\mathbf{X}(0:N)) \geq g_j(\mathbf{Y}(0:N))$,

e) if $\mathbf{X}(N) \geq \mathbf{Y}$, then $g_N(\mathbf{X}(0:N)) \geq \Gamma(\mathbf{Y})$,

f) if $\mathbf{X} \geq \mathbf{Y}$, then $\Gamma(\mathbf{X}) \geq \Gamma(\mathbf{Y})$.

PROOF.

a) As the operator $\Phi_j(\widehat{\mathbf{L}}(j), \mathbf{X}(0:N))$ is convex and quadratic in $\widehat{\mathbf{L}}(j)$ and the matrices $\mathbf{X}(0:N)$, $\widehat{\mathbf{R}}(0:N)$ are positive semidefinite, the minimizer of (B.6) with respect to $\widehat{\mathbf{L}}(j)$ is obtained by

$$\frac{\partial \Phi_j(\widehat{\mathbf{L}}(j), \mathbf{X}(0:N))}{\partial \widehat{\mathbf{L}}(j)} = 2 \cdot \left(\sum_{i=0}^{N} p(j,i) \widehat{\mathbf{R}}(i) \right) \widehat{\mathbf{L}}(j)$$
$$+ 2 \cdot \sum_{i=0}^{N} p(j,i) \left(\widehat{\mathbf{B}}(i)^\top \mathbf{X}(i) \widehat{\mathbf{B}}(i) \widehat{\mathbf{L}}(j) + \widehat{\mathbf{B}}(i)^\top \mathbf{X}(i) \widehat{\mathbf{A}}(i) \right) \stackrel{!}{=} 0 .$$

Solving for $\widehat{\mathbf{L}}(j)$ produces $\widehat{\mathbf{L}}(j) = L_j^{\Phi}(\mathbf{X}(0:N))$.

b) This follows from Lemma B.2 a) by substitution.

c) This fact is a direct implication of Lemma B.2 b).

d) $$\begin{aligned} g_j(\mathbf{Y}(0:N)) &= \Phi_j(L_j^{\Phi}(\mathbf{Y}(0:N)), \mathbf{Y}(0:N)) \\ &\leq \Phi_j(L_j^{\Phi}(\mathbf{X}(0:N)), \mathbf{Y}(0:N)) \\ &\leq \Phi_j(L_j^{\Phi}(\mathbf{X}(0:N)), \mathbf{X}(0:N)) = g_j(\mathbf{X}(0:N)) \ . \end{aligned}$$

e) With $\mathbf{X}(0:N) \geq (\mathbf{0},\ldots,\mathbf{0},\mathbf{Y})$ and Lemma B.2 d) it follows that

$$\begin{aligned} g_N(\mathbf{X}(0:N)) &\geq g_N(\mathbf{0},\ldots,\mathbf{0},\mathbf{Y}) \\ &= \Phi_N(L_N^{\Phi}(\mathbf{0},\ldots,\mathbf{0},\mathbf{Y}),(\mathbf{0},\ldots,\mathbf{0},\mathbf{Y})) \\ &= \Gamma(\mathbf{Y}) \ . \end{aligned}$$

f) This follows from Lemma B.2 d) with

$$\begin{aligned} \mathbf{X}(0:N) &= (\mathbf{0},\ldots,\mathbf{0},\mathbf{X}) \\ \mathbf{Y}(0:N) &= (\mathbf{0},\ldots,\mathbf{0},\mathbf{Y}) \ . \end{aligned}$$

□

Lemma B.3 Consider the operators

$$\mathcal{L}_j(\mathbf{Y}(0:N)) = \sum_{i=0}^{N} p(j,i) \left(\widehat{\mathbf{A}}(i) + \widehat{\mathbf{B}}(i)\mathbf{L}(j)\right)^{\top} \mathbf{Y}(i) \left(\widehat{\mathbf{A}}(i) + \widehat{\mathbf{B}}(i)\mathbf{L}(j)\right)$$

and

$$\mathcal{L}(\mathbf{Y}(0:N)) = (\mathcal{L}_0(\mathbf{Y}(0:N)),\ldots,\mathcal{L}_N(\mathbf{Y}(0:N))) \ ,$$

and assume that there exist $N+1$ matrices $\overline{\mathbf{Y}}(0:N) > 0$ such that $\overline{\mathbf{Y}}(0:N) > \mathcal{L}(\overline{\mathbf{Y}}(0:N))$, then

a) it holds for the sequence $\mathbf{M}_{k+1}(0:N) = \mathcal{L}(\mathbf{M}_k(0:N))$ initialized with $\mathbf{M}_0(0:N) \geq 0$ that $\lim_{k\to\infty} \mathbf{M}_k(0:N) = 0$, and

b) the sequence $\mathbf{M}_{k+1}(0:N) = \mathcal{L}(\mathbf{M}_k(0:N)) + (\mathbf{S}(0),\cdots,\mathbf{S}(N))$ initialized with $\mathbf{M}_0(0:N) \geq 0$ is bounded for all $\mathbf{S}(0:N) \geq 0$.

Proof.

a) Choose $0 \leq m$ such that $\mathbf{M}_0(0:N) \leq m\overline{\mathbf{Y}}(0:N)$. Furthermore, choose $0 < r < 1$ such that $\mathcal{L}(\overline{\mathbf{Y}}(0:N)) < r\overline{\mathbf{Y}}(0:N)$ and consider the sequence $\mathbf{N}_{k+1}(0:N) = \mathcal{L}(\mathbf{N}_k(0:N))$ initialized with $\mathbf{N}_0(0:N) = m\overline{\mathbf{Y}}(0:N)$. Then,

$$0 \leq \mathbf{M}_{k+1}(0:N) \leq \mathbf{N}_{k+1}(0:N) \leq mr^{(k+1)}\overline{\mathbf{Y}}(0:N) \ ,$$

since, first, $\mathbf{Y}(0:N) \geq 0$ implies that $\mathcal{L}(\mathbf{Y}(0:N)) \geq 0$ and, second, if $\mathbf{Y}(0:N) \geq \mathbf{X}(0:N)$ then $\mathcal{L}(\mathbf{Y}(0:N)) \geq \mathcal{L}(\mathbf{X}(0:N))$. Taking the limit $k \to \infty$ justifies the proposition.

b) Choose $0 \leq s$ such that $\mathbf{S}(0:N) \leq s\overline{\mathbf{Y}}(0:N)$. Consider the sequence $\mathbf{S}_{k+1}(0:N) = \mathcal{L}(\mathbf{S}_k(0:N))$ initialized with $\mathbf{S}_0(0:N) = \mathbf{S}(0:N)$ and the sequence $\mathbf{N}_{k+1}(0:N) = \mathcal{L}(\mathbf{N}_k(0:N))$ initialized with $\mathbf{N}_0(0:N) = \mathbf{M}_0(0:N)$. Then,

$$\mathbf{M}_{k+1}(0:N) = \mathbf{N}_{k+1}(0:N) + \sum_{t=0}^{k} \mathbf{S}_t(0:N)$$

and using Lemma B.3 a) it follows with $0 \leq m_N, m_U$ and with $0 < r_N, r_U < 1$ that

$$\begin{aligned}\mathbf{M}_{k+1}(0:N) &\leq m_N \cdot r_N^{k+1}\overline{\mathbf{Y}}(0:N) + \sum_{t=0}^{k} m_U \cdot r_U^t \overline{\mathbf{Y}}(0:N) \\ &\leq \left(m_N + \frac{m_U}{1-r}\right)\overline{\mathbf{Y}}(0:N)\,.\end{aligned}$$

□

B.1. Proof of Theorem 3.3 (LRAC-stability)

a) The proof is given in [118, Theorem 2]. Note that the corresponding conditions are satisfied due to Assumption 3.6 b), d), and e).

b) Consider the sequence $\mathbf{X}_{k+1}(0:N) = g(\mathbf{X}_k(0:N))$ defined in (B.1). Using Lemma B.2 c), it holds that

$$\begin{aligned}\mathbf{X}_{k+1}(0:N) = g(\mathbf{X}_k(0:N)) &\leq \Phi(\widehat{\mathbf{L}}(0:N), \mathbf{X}_k(0:N)) \\ &= \mathcal{L}(\mathbf{X}_k(0:N)) + (\mathbf{S}(0), \ldots, \mathbf{S}(N))\,,\end{aligned}$$

where

$$\mathbf{S}(j) \stackrel{\text{def}}{=} \sum_{i=0}^{N} p(j,i)\left(\widehat{\mathbf{Q}}(i) + (\widehat{\mathbf{L}}(j))^{\top}\widehat{\mathbf{R}}(i)\widehat{\mathbf{L}}(j)\right)\,,$$

$$\Phi(\widehat{\mathbf{L}}(0:N), \mathbf{X}_k(0:N)) \stackrel{\text{def}}{=} (\Phi_0(\widehat{\mathbf{L}}(0), \mathbf{X}_k(0:N)), \ldots,$$
$$\Phi_N(\widehat{\mathbf{L}}(N), \mathbf{X}_k(0:N)) \ ,$$

and $\mathcal{L}(\mathbf{X}\,(0:N))$ as defined in Lemma B.3. As it is assumed that $\mathbf{X}(0:N) > \mathcal{L}(\mathbf{X}(0:N))$, the condition of Lemma B.3 is satisfied so that it follows according to Lemma B.3 b) that the sequence $\mathbf{X}_{k+1}(0:N) = g(\mathbf{X}_k(0:N))$ is bounded. Finally, this combined with Lemma B.1 a) reveals that the system is LRAC-stable.

B.2. Proof of Theorem 3.4 (LRAC-instability)

a) The proof is given in [118, Theorem 4]. Note that the corresponding conditions are satisfied due to Assumption 3.6 b), d), and e).

b) Consider the sequences $\mathbf{X}_{k+1}(0:N) = g(\mathbf{X}_k(0:N))$ initialized with $\mathbf{X}_0(0:N) = \mathbf{0}$ and the sequence $\mathbf{N}_{k+1} = \Gamma(\mathbf{N}_k)$ with $\mathbf{N}_0 = \mathbf{0}$. Then, $\mathbf{X}_1(0:N) \geq 0$ and $\mathbf{X}_1(N) \geq \mathbf{N}_1$. It follows from Lemma B.2 d) that $\mathbf{X}_k(N) \geq \mathbf{N}_k$. This implies that $\mathbf{X}_k(N)$ is lower bounded by $\mathbf{N}_k$.

Moreover, if $\mathbf{N}_{k+1} = \Gamma(\mathbf{N}_k)$ converges, $\overline{\mathbf{N}} = \lim_{k\to\infty} \mathbf{N}_k$ has to be a fixed point of $\Gamma(\cdot)$ since $\Gamma(\cdot)$ is a continuous operator. Using (B.8), the fixed point equation $\overline{\mathbf{N}} = \Gamma\big(\overline{\mathbf{N}}\big)$ is $\overline{\mathbf{N}} = \overline{\mathbf{Q}} + \overline{\mathbf{A}}^\top \overline{\mathbf{N}}\overline{\mathbf{A}}$, with

$$\overline{\mathbf{Q}} = \sum_{i=0}^{N} p(j,i)\widehat{\mathbf{Q}}(i) + p(N,N)\overline{\mathbf{L}}^\top \widehat{\mathbf{R}}(i)\overline{\mathbf{L}} \ ,$$
$$\overline{\mathbf{A}} = \sqrt{p(N,N)}\left(\widehat{\mathbf{A}}(N) + \widehat{\mathbf{B}}(N)\overline{\mathbf{L}}\right).$$

From the observability of $(\mathbf{A}, \mathbf{Q}^{1/2})$ (Assumption 3.6 c), it follows by means of the Belovich-Popov-Hautus test that the pair $(\widehat{\mathbf{A}}(N), (\sum_{i=0}^{N} p(j,i)\widehat{\mathbf{Q}}(i))^{1/2})$ is observable. Consequently, $(\overline{\mathbf{A}}, \overline{\mathbf{Q}}^{1/2})$ is also observable. In addition, since $\overline{\mathbf{Q}} \geq 0$, it follows according to Lyapunov theory that if $\max|\mathrm{eig}(\overline{\mathbf{A}})| > 1$, then there exists no positive semidefinite solution to the fixed point equation. In addition, noting that

$$\widehat{\mathbf{A}}(N) + \widehat{\mathbf{B}}(N)\overline{\mathbf{L}} = \widehat{\mathbf{A}}(N) + \widehat{\mathbf{B}}(N) \begin{bmatrix} \overline{\mathbf{L}}_{11} & \overline{\mathbf{L}}_{12} \\ \overline{\mathbf{L}}_{21} & \overline{\mathbf{L}}_{22} \end{bmatrix} = \begin{bmatrix} \mathbf{A} & \mathbf{0} \\ \overline{\mathbf{L}}_{21} & \overline{\mathbf{L}}_{22} \\ \mathbf{0} & \mathbf{F} \end{bmatrix} ,$$

it turns out that $\mathrm{eig}(\mathbf{A}) \subset \mathrm{eig}(\overline{\mathbf{A}})$. Therefore, the sequence $\mathbf{N}_k$ has no fixed point if

$$\sqrt{p(N,N)}\,\max|\mathrm{eig}(\mathbf{A})| > 1 \; . \tag{B.9}$$

If condition (B.9) holds, the sequence $\mathbf{N}_k$ diverges because the sequence does not converge to a fixed point, on the one hand, and the sequence is monotonically increasing (Lemma B.2 f), on the other hand. Noting that $\mathbf{X}_k(N) \geq \mathbf{N}_k$, the system is LRAC-unstable according to Lemma B.1 a).

B.3. Proof of Theorem 3.5

Consider the sequence $\mathbf{X}_{k+1}(0:N) = g(\mathbf{X}_k(0:N))$ defined in (B.1) and initialized with $\mathbf{X}_0(0:N) = \mathbf{0}$. According to Lemma B.2 d), the sequence increases monotonically, and according to Theorem B.1 b), the sequence is bounded from above. Hence, the sequence converges. In addition, as $g(\cdot)$ is a continuous function, its limit has to be the fixed point $\overline{\mathbf{K}}(0:N) = g\big(\overline{\mathbf{K}}(0:N)\big)$.

C. Proof of Theorem 4.1 (Optimal Tracking Controller)

To prove Theorem 4.1 , we use the same technique as in Appendix A. The cost function is reformulated in terms of the augmented state. In this way, the Markov property of the system is restored and we can apply the dynamic programming algorithm. By doing so, the general optimization problem is separated into several recursively coupled subproblems that are solved analytically.

Using the sequence-based NCS model (3.3) and the matrices (4.10) and (4.11), we can express the minimum cumulative costs in Problem 4.1 as

$$
\begin{aligned}
C_{0\to K}^{Trk^*} = \min_{\underline{U}_{0:K-1}} \mathrm{E}\Bigg\{ & (\underline{z}_K^{Ref})^\top \mathbf{Q}_K \underline{z}_K^{Ref} + \underline{\xi}_K^\top \widehat{\mathbf{Q}}_K \underline{\xi}_K - 2(\underline{z}_K^{Ref})^\top \overline{\mathbf{Q}}_K \underline{\xi}_K \\
& + \sum_{k=0}^{K-1} \Big[(\underline{z}_k^{Ref})^\top \mathbf{Q}_k \underline{z}_k^{Ref} + \underline{\xi}_k^\top \widehat{\mathbf{Q}}_k(\theta_k) \underline{\xi}_k - 2(\underline{z}_k^{Ref})^\top \overline{\mathbf{Q}}_k \underline{\xi}_k \\
& + \underline{U}_k^\top \widehat{\mathbf{R}}_k(\theta_k) \underline{U}_k \Big] \Big| \mathcal{I}_0 \Bigg\} \ .
\end{aligned}
$$

According to dynamic programming theory, the relation with the minimum costs-to-go holds

$$
J_0^{Trk^*} = C_{0\to K}^{Trk^*} \ , \tag{C.1}
$$

where $J_0^{Trk^*}$ is given by the recursion

$$
\begin{aligned}
J_k^{Trk^*} = \min_{\underline{U}_k} \mathrm{E}\Big\{ & (\underline{z}_k^{Ref})^\top \mathbf{Q}_k \underline{z}_k^{Ref} + \underline{\xi}_k^\top \widehat{\mathbf{Q}}_k(\theta_k) \underline{\xi}_k - 2(\underline{z}_k^{Ref})^\top \overline{\mathbf{Q}}_k \underline{\xi}_k \\
& + \underline{U}_k^\top \widehat{\mathbf{R}}_k(\theta_k) \underline{U}_k + J_{k+1}^{Trk^*} \Big| \mathcal{I}_k \Big\} ,
\end{aligned} \tag{C.2}
$$

that is initialized at K with

$$J_K^{Trk^*} = \mathrm{E}\left\{(\underline{z}_K^{Ref})^\top \mathbf{Q}_K \underline{z}_K^{Ref} + \underline{\xi}_K^\top \widehat{\mathbf{Q}}_K \underline{\xi}_K - 2(\underline{z}_K^{Ref})^\top \overline{\mathbf{Q}}_K \underline{\xi}_K \middle| \mathcal{I}_K\right\} . \tag{C.3}$$

Using (C.3) and the definitions in (4.11), the expected costs-to-go at time step K are

$$\begin{aligned} J_K^{Trk^*} &= \mathrm{E}\left\{(\underline{z}_K^{Ref})^\top \mathbf{Q}_K \underline{z}_K^{Ref} + \underline{\xi}_K^\top \widehat{\mathbf{Q}}_K(\theta_K) \underline{\xi}_K - 2(\underline{z}_K^{Ref})^\top \overline{\mathbf{Q}}_K \underline{\xi}_K \middle| \mathcal{I}_K\right\} \\ &= s_K + \mathrm{E}\left\{\underline{\xi}_K^\top \mathbf{K}_K \underline{\xi}_K \middle| \mathcal{I}_K\right\} - 2 \cdot \underline{\sigma}_K^\top \mathrm{E}\left\{\underline{\xi}_K \middle| \mathcal{I}_K\right\} . \end{aligned} \tag{C.4}$$

Plugging (C.4) into (C.3), it yields the minimal costs-to-go at time step $K-1$

$$\begin{aligned} J_{K-1}^{Trk^*} = \min_{\underline{U}_k} \Bigg[& \mathrm{E}\Big\{\underline{\xi}_{K-1}^\top \Big(\widehat{\mathbf{Q}}_{K-1}(\theta_{K-1}) \\ & + \widehat{\mathbf{A}}_{K-1}(\theta_{K-1})^\top \mathbf{K}_K \widehat{\mathbf{A}}_{K-1}(\theta_{K-1})\Big) \underline{\xi}_{K-1} \Big| \mathcal{I}_{K-1}\Big\} \\ & + \underline{U}_{K-1}^\top \mathrm{E}\left\{\widehat{\mathbf{R}}_{K-1}(\theta_{K-1}) + \widehat{\mathbf{B}}_{K-1}(\theta_{K-1})^\top \mathbf{K}_K \widehat{\mathbf{B}}_{K-1}(\theta_{K-1}) \middle| \mathcal{I}_{K-1}\right\} \underline{U}_{K-1} \\ & + 2 \cdot \mathrm{E}\left\{\underline{\xi}_{K-1}^\top \middle| \mathcal{I}_{K-1}\right\} \mathrm{E}\left\{\widehat{\mathbf{A}}_{K-1}(\theta_{K-1})^\top \mathbf{K}_K \widehat{\mathbf{B}}_{K-1}(\theta_{K-1}) \middle| \mathcal{I}_{K-1}\right\} \underline{U}_{K-1} \\ & - 2 \cdot \left[\underline{\sigma}_K^\top \mathrm{E}\left\{\widehat{\mathbf{A}}_{K-1}(\theta_{K-1}) \middle| \mathcal{I}_{K-1}\right\} + (\underline{z}_{K-1}^{Ref})^\top \overline{\mathbf{Q}}_{K-1}\right] \\ & \cdot \mathrm{E}\left\{\underline{\xi}_{K-1} \middle| \mathcal{I}_{K-1}\right\} - 2 \cdot \underline{\sigma}_K^\top \mathrm{E}\left\{\widehat{\mathbf{B}}_{K-1}(\theta_{K-1}) \middle| \mathcal{I}_{K-1}\right\} \underline{U}_{K-1} \\ & + (\underline{z}_{K-1}^{Ref})^\top \mathbf{Q}_{K-1} \underline{z}_{K-1}^{Ref} + \mathrm{E}\left\{\underline{\widehat{w}}_{K-1}^\top \mathbf{K}_K \underline{\widehat{w}}_{K-1} \middle| \mathcal{I}_{K-1}\right\} + s_K \Bigg] . \end{aligned} \tag{C.5}$$

Differentiation of (C.5) and solving for $\underline{U}_{K-1}$ yields the minimizer

$$\begin{aligned} \underline{U}_{K-1} = - & \left(\mathrm{E}\left\{\widehat{\mathbf{R}}_{K-1}(\theta_{K-1}) + \widehat{\mathbf{B}}_{K-1}(\theta_{K-1})^\top \mathbf{K}_K \widehat{\mathbf{B}}_{K-1}(\theta_{K-1}) \middle| \mathcal{I}_{K-1}\right\}\right)^\dagger \\ & \cdot \Big[\mathrm{E}\left\{\widehat{\mathbf{B}}_{K-1}(\theta_{K-1})^\top \mathbf{K}_K \widehat{\mathbf{A}}_{K-1}(\theta_{K-1}) \middle| \mathcal{I}_{K-1}\right\} \mathrm{E}\left\{\underline{\xi}_{K-1} \middle| \mathcal{I}_{K-1}\right\} \\ & - \mathrm{E}\left\{\widehat{\mathbf{B}}_{K-1}(\theta_{K-1})^\top \middle| \mathcal{I}_{K-1}\right\} \underline{\sigma}_K \Big] . \end{aligned} \tag{C.6}$$

Using (C.6) in (C.5) with definitions (4.7) – (4.13), the minimal costs-to-go at time step $K-1$ are

$$J_{K-1}^{Trk^*} = \mathrm{E}\left\{\underline{\xi}_{K-1}^\top \mathbf{K}_{K-1}\underline{\xi}_{K-1}\middle|\mathcal{I}_{K-1}\right\} - 2\cdot\underline{\sigma}_{K-1}^\top \mathrm{E}\left\{\underline{\xi}_{K-1}\middle|\mathcal{I}_{K-1}\right\} + s_{K-1}$$
$$+\mathrm{E}\left\{\underline{\widehat{w}}_{K-1}^\top \mathbf{K}_K \underline{\widehat{w}}_{K-1}\middle|\mathcal{I}_{K-1}\right\} + \mathrm{E}\left\{\underline{e}_{K-1}^\top \mathbf{P}_{K-1}\underline{e}_{K-1}\middle|\mathcal{I}_{K-1}\right\}\ . \tag{C.7}$$

With (C.7), we can calculate the expected costs-to-go at time step $K-2$ by using (C.3) such that

$$J_{K-2}^{Trk^*} = \min_{\underline{U}_{K-2}}\Bigg[\underline{U}_{K-2}^\top \mathrm{E}\Big\{\widehat{\mathbf{R}}_{K-2}(\theta_{K-2}) \tag{C.8}$$
$$+\ \widehat{\mathbf{B}}_{K-2}(\theta_{K-2})^\top \mathbf{K}_{K-1}\widehat{\mathbf{B}}_{K-2}(\theta_{K-2})\Big|\mathcal{I}_{K-2}\Big\}\underline{U}_{K-2}$$
$$+2\cdot\mathrm{E}\left\{\underline{\xi}_{K-2}^\top\middle|\mathcal{I}_{K-2}\right\}\mathrm{E}\left\{\widehat{\mathbf{A}}_{K-2}(\theta_{K-2})^\top\mathbf{K}_{K-1}\widehat{\mathbf{B}}_{K-2}(\theta_{K-2})\middle|\mathcal{I}_{K-2}\right\}\underline{U}_{K-2}$$
$$-2\cdot\mathrm{E}\left\{\underline{\sigma}_{K-1}^\top\widehat{\mathbf{B}}_{K-2}(\theta_{K-2})\middle|\mathcal{I}_{K-2}\right\}\underline{U}_{K-2}\Bigg]$$
$$-2\cdot\left[\mathrm{E}\left\{\underline{\sigma}_{K-1}^\top\widehat{\mathbf{A}}_{K-2}(\theta_{K-2})\middle|\mathcal{I}_{K-2}\right\} + (\underline{z}_{K-2}^{Ref})^\top\overline{\mathbf{Q}}_{K-2}\right]\cdot\mathrm{E}\left\{\underline{\xi}_{K-2}\middle|\mathcal{I}_{K-2}\right\}$$
$$+\mathrm{E}\left\{\underline{\xi}_{K-2}^\top\left(\widehat{\mathbf{Q}}_{K-2}(\theta_{K-2})+\widehat{\mathbf{A}}_{K-2}(\theta_{K-2})^\top\mathbf{K}_{K-1}\widehat{\mathbf{A}}_{K-2}(\theta_{K-2})\right)\underline{\xi}_{K-2}\middle|\mathcal{I}_{K-2}\right\}$$
$$+\mathrm{E}\left\{\underline{e}_{K-1}^\top\mathbf{P}_{K-1}\underline{e}_{K-1}\middle|\mathcal{I}_{K-2}\right\} + (\underline{z}_{K-2}^{Ref})^\top\mathbf{Q}_{K-2}\underline{z}_{K-2}^{Ref}$$
$$+\mathrm{E}\left\{s_{K-1}\middle|\mathcal{I}_{K-2}\right\} + \sum_{k=K-2}^{K-1}\mathrm{E}\left\{\underline{\widehat{w}}_k^\top\mathbf{K}_{k+1}\underline{\widehat{w}}_k\middle|\mathcal{I}_{K-2}\right\}\ .$$

As in the S-LQG case, the term $\mathrm{E}\left\{\underline{e}_{K-1}^\top\mathbf{P}_{K-1}\underline{e}_{K-1}\middle|\mathcal{I}_{K-2}\right\}$ is independent of $\underline{U}_{K-2}$, which is justified by Lemma A.3. The minimal expected costs-to-go $J_{K-2}^{Trk^*}$ can be calculated similarly to $J_{K-1}^{Trk^*}$ as the structure of (C.7) and (C.8) is identical. Proceeding with the optimization, we therefore get to the solution stated by Theorem 4.1 .

D. Proof of Lemma 5.1 (Event-Triggered S-LQG)

The individual costs (5.14) and (5.15) can be computed in the same way as the costs $C^*_{0\to K}$ (3.33) in Appendix A. Therefore, replacing the term $p(i,j)$ by $p_l^{Evt[k]}(i,j)$ and exchanging $\mathcal{I}_k$ with $\mathcal{S}_l^{[k]}$ in Appendix A, we obtain

$$\begin{aligned} C_{k\to K}^{Send} = \mathrm{E}\left\{\underline{\xi}_k^\top \mathbf{K}_k^{[k]} \underline{\xi}_k \middle| \mathcal{S}_k^{[k]}\right\} + \sum_{i=k}^{K-1} \mathrm{E}\left\{\underline{\widehat{w}}_i^\top \mathbf{K}_i \underline{\widehat{w}}_i \middle| \mathcal{S}_k^{[k]}\right\} + \sum_{i=k}^{K-1} S_i \\ + \sum_{i=k+1}^{K-1} \mathrm{E}\left\{\left(e_i^{Send}\right)^\top \mathbf{P}_i e_i^{Send} \middle| \mathcal{S}_k^{[k]}\right\} \ , \end{aligned} \tag{D.1}$$

with

$$e_i^{Send} \stackrel{\text{def}}{=} \underline{\xi}_i - \mathrm{E}\left\{\underline{\xi}_i \middle| \mathcal{S}_i^{[k]}\right\} \ .$$

The costs for not sending the sequence, $C_{k\to K}^{NotSend}$, can be calculated analogously by replacing $\mathcal{I}_k$ with $\mathcal{N}_l^{[k]}$ instead. The result is

$$\begin{aligned} C_{k\to K}^{NotSend} = \ & \mathrm{E}\left\{\underline{\xi}_k^\top \mathbf{K}_k^{[k]} \underline{\xi}_k \middle| \mathcal{N}_k^{[k]}\right\} + \sum_{i=k}^{K-1} \mathrm{E}\left\{\underline{\widehat{w}}_i^\top \mathbf{K}_i^{[k]} \underline{\widehat{w}}_i \middle| \mathcal{N}_k^{[k]}\right\} \\ & + \sum_{i=k+1}^{K-1} S_i + \sum_{i=k+1}^{K-1} \mathrm{E}\left\{\left(e_i^{NotSend}\right)^\top \mathbf{P}_i e_i^{NotSend} \middle| \mathcal{N}_k^{[k]}\right\}, \end{aligned} \tag{D.2}$$

with

$$e_i^{NotSend} \stackrel{\text{def}}{=} \left(\underline{\xi}_i - \mathrm{E}\left\{\underline{\xi}_i \middle| \mathcal{N}_i^{[k]}\right\}\right) \ .$$

It holds for the difference $C_{k\to K}^{Send} - C_{k\to K}^{NotSend}$ of these costs

$$C_{k\to K}^{Send} - C_{k\to K}^{NotSend} = \Delta_{k:K}^{Con} + \Delta_{k:K}^{Est} \ , \tag{D.3}$$

with

$$\begin{aligned}\Delta^{Con}_{k:K} &\stackrel{\text{def}}{=} \mathrm{E}\left\{\underline{\xi}_k^\top \mathbf{K}_k^{[k]} \underline{\xi}_k \middle| \mathcal{S}_k^{[k]}\right\} - \mathrm{E}\left\{\underline{\xi}_k^\top \mathbf{K}_k^{[k]} \underline{\xi}_k \middle| \mathcal{N}_k^{[k]}\right\} ,\\ &+ \sum_{i=k}^{K-1} \mathrm{E}\left\{\underline{\widehat{w}}_i^\top \mathbf{K}_i \underline{\widehat{w}}_i \middle| \mathcal{S}_k^{[k]}\right\} - \sum_{i=k}^{K-1} \mathrm{E}\left\{\underline{\widehat{w}}_i^\top \mathbf{K}_i \underline{\widehat{w}}_i \middle| \mathcal{N}_k^{[k]}\right\} + S ,\\ \Delta^{Est}_{k:K} &\stackrel{\text{def}}{=} \sum_{i=k}^{K-1} \mathrm{E}\left\{\left(e_i^{Send}\right)^\top \mathbf{P}_i e_i^{Send} \middle| \mathcal{S}_k^{[k]}\right\}\\ &- \sum_{i=k}^{K-1} \mathrm{E}\left\{\left(e_i^{NotSend}\right)^\top \mathbf{P}_i e_i^{NotSend} \middle| \mathcal{N}_k^{[k]}\right\} .\end{aligned}$$

Furthermore, Lemma A.3 also holds for the information structures $\mathcal{S}_k^{[k]}$ and $\mathcal{N}_k^{[k]}$ due to Assumption 3.5. This implies that the expressions

$$\mathrm{E}\left\{\left(e_i^{Send}\right)^\top \mathbf{P}_i e_i^{Send} \middle| \mathcal{S}_k^{[k]}\right\} \tag{D.4}$$

and

$$\mathrm{E}\left\{\left(e_i^{NotSend}\right)^\top \mathbf{P}_i e_i^{NotSend} \middle| \mathcal{N}_k^{[k]}\right\} \tag{D.5}$$

are independent of $\underline{U}_{0:k-1}$ and $\underline{U}_k^{cand}$ for all $i \in \{k, \ldots, K\}$ and, hence,

$$\Delta^{Est}_{k:K} = 0 .$$

In analogy to Corollary 5.1, we can observe that the approximation of the transition matrix for the case that $\underline{U}_k^{cand}$ is not sent to the actuator also converges after N time steps towards the transition matrix of the S-LQG. Therefore, it holds that

$$\sum_{i=k+N}^{K-1} \mathrm{E}\left\{\underline{\widehat{w}}_i^\top \mathbf{K}_i^{[k]} \underline{\widehat{w}}_i \middle| \mathcal{S}_k^{[k]}\right\} - \sum_{i=k+N}^{K-1} \mathrm{E}\left\{\underline{\widehat{w}}_i^\top \mathbf{K}_i^{[k]} \underline{\widehat{w}}_i \middle| \mathcal{N}_k^{[k]}\right\} = 0 ,$$

Using this result in (D.3), finally results in

$$\begin{aligned}C^{Send}_{k\to K} - C^{NotSend}_{k\to K} &= \mathrm{E}\left\{\underline{\xi}_k^\top \mathbf{K}_k^{[k]} \underline{\xi}_k \middle| \mathcal{S}_k^{[k]}\right\} - \mathrm{E}\left\{\underline{\xi}_k^\top \mathbf{K}_k^{[k]} \underline{\xi}_k \middle| \mathcal{N}_k^{[k]}\right\}\\ &+ \sum_{i=k}^{k+N} \mathrm{E}\left\{\underline{\widehat{w}}_i^\top \mathbf{K}_i \underline{\widehat{w}}_i \middle| \mathcal{S}_k^{[k]}\right\} - \sum_{i=k}^{k+N} \mathrm{E}\left\{\underline{\widehat{w}}_i^\top \mathbf{K}_i \underline{\widehat{w}}_i \middle| \mathcal{N}_k^{[k]}\right\} + S_k ,\end{aligned}$$

concluding the proof.

List of Figures

Bibliography

[1] G. Alldredge, M. Branicky, and V. Liberatore, "Play-Back Buffers in Networked Control Systems: Evaluation and Design," in *Proc. of the American Control Conference*, pp. 3106–3113, 2008.

[2] B. D. O. Anderson and J. B. Moore, *Optimal Control: Linear Quadratic Methods*, vol. 1, Englewood Cliffs, New Jersey: Prentice Hall, 1990.

[3] C. W. Anderson, "Learning to Control an Inverted Pendulum using Neural Networks," *IEEE Control Systems Magazine*, vol. 9, no. 3, pp. 31 –37, 1989.

[4] P. Antsaklis and J. Baillieul, "Special Issue on Technology of Networked Control Systems," *Proc. of the IEEE*, vol. 95, no. 1, pp. 5–8, 2007.

[5] D. Antunes and W. P. M. H. Heemels, "Rollout Event-Triggered Control: Beyond Periodic Control Performance," 2014, (accepted).

[6] P. Arcara and C. Melchiorri, "Control Schemes for Teleoperation with Time Delay: A Comparative Study," *Robotics and Autonomous Systems*, vol. 38, no. 1, pp. 49 – 64, 2002.

[7] K. J. Åström and B. M. Bernhardsson, "Comparison of Riemann and Lebesgue Sampling for First Order Stochastic Systems," in *Proc. of the IEEE Conference on Decision and Control*, vol. 2, pp. 2011–2016, 2002.

[8] K. J. Åström and P. R. Kumar, "Control: A Perspective," *Automatica*, vol. 50, no. 1, pp. 3 – 43, 2014.

[9] J. Baillieul and P. J. Antsaklis, "Control and Communication Challenges in Networked Real-Time Systems," *Proc. of the IEEE*, vol. 95, no. 1, pp. 9–28, 2007.

[10] Y. Bar-Shalom and E. Tse, "Dual Effect, Certainty Equivalence, and Separation in Stochastic Control," *IEEE Transactions on Automatic Control*, vol. 19, no. 5, pp. 494–500, 1974.

[11] A. Bemporad, "Predictive Control of Teleoperated Constrained Systems with Unbounded Communication Delays," in *Proc. of the IEEE Conference on Decision and Control*, vol. 2, pp. 2133–2138, 1998.

[12] A. Bemporad, W. P. M. H. Heemels, and M. Johansson, *Networked Control Systems*, vol. 406 of *Lecture Notes in Control and Information Sciences*, London: Springer, 2010.

[13] D. Bertsekas, "Dynamic Programming and Suboptimal Control: A Survey from ADP to MPC," *European Journal of Control*, vol. 11, no. 4-5, pp. 310–334, 2005.

[14] D. P. Bertsekas, *Dynamic Programming and Optimal Control*, vol. 1, Belmont, Massachusetts: Athena Scientific, 2nd edn., 2000.

[15] D. P. Bertsekas, *Dynamic Programming and Optimal Control*, vol. 2, Belmont, Massachusetts: Athena Scientific, 2nd edn., 2000.

[16] R. Blind, U. Münz, and F. Allgöwer, "Modeling, Analysis, and Design of Networked Control Systems Using Jump Linear Systems," *at - Automatisierungstechnik*, vol. 56, no. 1, p. 20, 2008.

[17] S. Boyd, L. E. Ghaoui, E. Feron, and V. Balakrishnan, *Linear Matrix Inequalities in System and Control Theory*, vol. 15, Philadelphia: Society for Industrial and Applied Mathematics, 1994.

[18] M. Bregulla, W. Feucht, J. Koch, and J. W. u. J. W. G. de Mür, M. Schade, *Wireless Solutions in Automation - Overview and Decision Support*Tech. rep., ZVEI - German Electrical and Electronic Manufacuters' Association, Frankfurt am Main, 2011.

[19] E. F. Camacho and C. Bordons, *Model Predictive Control*, London: Springer, 2007.

[20] CAN in Automation e.V., *http://www.can-cia.org*

[21] A. Casavola, E. Mosca, and M. Papini, "Predictive Teleoperation of Constrained Dynamic Systems via Internet-Like Channels," *IEEE Transactions on Control Systems Technology*, vol. 14, no. 4, pp. 681–694, 2006.

[22] A. Chaillet and A. Bicchi, "Delay Compensation in Packet-Switching Networked Controlled Systems," in *Proc. of the IEEE Conference on Decision and Control*, pp. 3620–3625, 2008.

[23] M. B. G. Cloosterman, N. van de Wouw, W. P. M. H. Heemels, and H. Nijmeijer, "Stability of Networked Control Systems With Uncertain Time-Varying Delays," *IEEE Transactions on Automatic Control*, vol. 54, no. 7, pp. 1575–1580, 2009.

[24] R. Cogill, "Event-Based Control using Quadratic Approximate Value Functions," in *Proc. of the IEEE Conference on Decision and Control*, pp. 5883–5888, 2009.

[25] D. Dzung, M. Naedele, T. P. von Hoff, and M. Crevatin, "Security for Industrial Communication Systems," *Proc. of the IEEE*, vol. 93, no. 6, pp. 1152–1177, 2005.

[26] N. Elia, "Remote Stabilization over Fading Channels," *Systems & Control Letters*, vol. 54, no. 3, pp. 237 – 249, 2005.

[27] E. O. Elliott, "Estimates of Error Rates for Codes on Burst-Noise Channels," *Bell System Technical Journal*, vol. 42, no. 5, pp. 1977–1997, 1963.

[28] M. Epstein, L. Shi, and R. M. Murray, "An Estimation Algorithm for a Class of Networked Control Systems Using UDP-Like Communication Schemes," in *Proc. of the IEEE Conference on Decision and Control*, pp. 5597–5603, 2006.

[29] EtherCAT Technology Group, *http://www.ethercat.org*

[30] S. Falasca, M. Gamba, L. Greco, A. Chaillet, and A. Bicchi, "Packet-based dynamic control for nonlinear networked systems," in *Proc. of the IEEE Conference on Decision and Control*, pp. 6597–6603, 2013.

[31] A. Ferrante and L. Ntogramatzidis, "The Generalised Discrete Algebraic Riccati Equation in Linear-Quadratic Optimal Control," *Automatica*, vol. 49, no. 2, pp. 471 – 478, 2013.

[32] Fieldbus Foundation, *http://www.fieldbus.org*

[33] R. Findeisen and P. Varutti, "Stabilizing Nonlinear Predictive Control over Nondeterministic Communication Networks," in *Nonlinear Model Predictive Control*, vol. 384 of *Lecture Notes in Control and Information Sciences*, pp. 167–179, Berlin: Springer, 2009.

[34] H. Gao and T. Chen, "Network-Based Output Tracking Control," *IEEE Transactions on Automatic Control*, vol. 53, no. 3, pp. 655–667, 2008.

[35] E. Garcia and P. Antsaklis, "Model-Based Event-Triggered Control for Systems With Quantization and Time-Varying Network Delays," *IEEE Transactions on Automatic Control*, vol. 58, no. 2, pp. 422–434, 2013.

[36] E. Garcia and P. Antsaklis, "Model-Based Event-Triggered Control for Systems With Quantization and Time-Varying Network Delays," *IEEE Transactions on Automatic Control*, vol. 58, no. 2, pp. 422–434, 2013.

[37] D. Georgiev and D. M. Tilbury, "Packet-Based Control: The H_2-Optimal Solution," *Automatica*, vol. 42, no. 1, pp. 137 – 144, 2006.

[38] E. N. Gilbert, "Capacity of a Burst-Noise Channel," *Bell System Technical Journal*, vol. 39, no. 5, pp. 1253–1265, 1960.

[39] F. Göktas, "Distributed Control of Systems over Communication Networks," Ph.D. thesis, Ph.D. Dissertation, University of Pennsylvania, 2000.

[40] T. Gommans, D. Antunes, T. Donkers, P. Tabuada, and W. P. M. H. Heemels, "Self-Triggered Linear Quadratic Control," *Automatica*, vol. 50, no. 4, pp. 1279 – 1287, 2014.

[41] L. Greco, A. Chaillet, and A. Bicchi, "Exploiting Packet Size in Uncertain Nonlinear Networked Control Systems," *Automatica*, vol. 48, no. 11, pp. 2801–2811, 2012.

[42] L. Grüne, J. Pannek, M. Seehafer, and K. Worthmann, "Analysis of Unconstrained Nonlinear MPC Schemes with Time Varying Control Horizon," *SIAM Journal on Control and Optimization*, vol. 48, no. 8, pp. 4938–4962, 2010.

[43] L. Grüne, J. Pannek, and K. Worthmann, "A Networked Unconstrained Nonlinear MPC Scheme," in *Proc. of the European Control Conference*, pp. 371–376, 2009.

[44] L. Grüne, J. Pannek, and K. Worthmann, "A Prediction Based Control Scheme for Networked Systems with Delays and Packet Dropouts," in *Proc. of the IEEE Conference on Decision and Control held jointly with the Chinese Control Conference*, pp. 537–542, 2009.

[45] V. Gupta, B. Sinopoli, S. Adlakha, A. Goldsmith, and R. Murray, "Receding Horizon Networked Control," in *In Proc. of the Annual Allerton Conference*, pp. 169–176, 2006.

[46] Y. Halevi and A. Ray, "Integrated Communication and Control Systems: Part I - Analysis," *Journal of Dynamic Systems, Measurement, and Control*, vol. 110, no. 4, pp. 367–373, 1988.

[47] A. Hassibi, S. P. Boyd, and J. P. How, "Control of Asynchronous Dynamical Systems with Rate Constraints on Events," in *Proc. of the IEEE Conference on Decision and Control*, vol. 2, pp. 1345–1351, 1999.

[48] W. P. M. H. Heemels and M. C. F. Donkers, "Model-Based Periodic Event-Triggered Control for Linear Systems," *Automatica*, vol. 49, no. 3, pp. 698 – 711, 2013.

[49] J. P. Hespanha, P. Naghshtabrizi, and Y. Xu, "A Survey of Recent Results in Networked Control Systems," *Proc. of the IEEE*, vol. 95, no. 1, pp. 138–162, 2007.

[50] K. Hikichi, H. Morino, I. Arimoto, K. Sezaki, and Y. Yasuda, "The Evaluation of Delay Jitter for Haptics Collaboration over the Internet," in *Proc. of the IEEE Global Telecommunications Conference (GLOBECOM)*, vol. 2, pp. 1492–1496, 2002.

[51] D. Hofstadter, *Gödel, Escher, Bach. An Eternal Golden Braid*, New York: Vintage Books, 1989.

[52] R. A. Horn and C. R. Johnson, *Matrix Analysis*, Cambridge University Press, 2 edn., 2012.

[53] W. Hu, G. Liu, and D. Rees, “Event-Driven Networked Predictive Control,” *IEEE Transactions on Industrial Electronics*, vol. 54, no. 3, pp. 1603–1613, 2007.

[54] W. Hu, G. Liu, and D. Rees, “Networked Predictive Control System with Data Compression,” in *IEEE Conference on Networking, Sensing and Control*, pp. 52–57, 2007.

[55] O. C. Imer, S. Yüksel, and T. Başar, “Optimal Control of LTI Systems over Unreliable Communication Links,” *Automatica*, vol. 42, no. 9, pp. 1429 – 1439, 2006.

[56] S. Johannessen, “Time Synchronization in a Local Area Network,” *IEEE Control Systems Magazine*, vol. 24, no. 2, pp. 61–69, 2004.

[57] M. Johansson and R. Jäntti, “Wireless Networking for Control: Technologies and Models,” in A. Bemporad, W. P. M. H. Heemels, and M. Johansson, editors, *Networked Control Systems*, vol. 406 of *Lecture Notes in Control and Information Sciences*, pp. 31–74, London: Springer, 2010.

[58] R. E. Kalman, “Contributions of the Theory of Optimal Control,” *Boletin de la Sociedad Matematica Mexicana*, vol. 5, no. 2, pp. 102–119, 1960.

[59] P. A. Kawka and A. G. Alleyne, “Stability and Performance of Packet-Based Feedback Control over a Markov Channel,” in *Proc. of the American Control Conference*, pp. 2807–2812, 2006.

[60] W. J. Kim, K. Ji, and A. Ambike, “Networked Real-Time Control Strategies Dealing with Stochastic Time Delays and Packet Losses,” in *Proc. of the American Control Conference*, pp. 621–626, 2005.

[61] Y. H. Kim, W. H. Kwon, and H. S. Park, “Stability and a Scheduling Method for Network-Based Control Systems,” in *Proc. of the IEEE International Conference on Industrial Electronics, Control, and Instrumentation (IECON)*, vol. 2, pp. 934–939, 1996.

[62] P. K. Kitanidis, “Unbiased Minimum-variance Linear State Estimation,” *Automatica*, vol. 23, no. 6, pp. 775 – 778, 1987.

[63] R. Langner, “Stuxnet: Dissecting a Cyberwarfare Weapon,” *IEEE Security & Privacy*, vol. 9, no. 3, pp. 49–51, 2011.

[64] D. Lehmann and J. Lunze, "Event-Based Control with Communication Delays and Packet Losses," *International Journal of Control*, vol. 85, no. 5, pp. 563–577, 2012.

[65] F. L. Lian, J. R. Moyne, and D. M. Tilbury, "Performance Evaluation of Control Networks: Ethernet, ControlNet, and DeviceNet," *IEEE Control Systems Magazine*, vol. 21, no. 1, pp. 66–83, 2001.

[66] F. L. Lian, W. Moyne, and D. Tilbury, "Network Design Consideration for Distributed Control Systems," *IEEE Transactions on Control Systems Technology*, vol. 10, no. 2, pp. 297–307, 2002.

[67] L. W. Liou and A. Ray, "A Stochastic Regulator for Integrated Communication and Control Systems: Part I-Formulation of Control Law," *ASME Journal of Dynamic Systems, Measurement and Control*, vol. 113, no. 4, pp. 604–611, 1991.

[68] L. Litz, O. Gabel, and I. Solihin, "NCS-Controllers for Ambient Intelligence Networks - Control Performance versus Control Effort," in *Proc. of the IEEE Conference on Decision and Control and the European Control Conference*, pp. 1571–1576, 2005.

[69] L. Litz, T. Gabriel, M. Groß, M., and O. Gabel, "Networked Control Systems (NCS) - Stand und Ausblick (Networked Control Systems (NCS) - State of the Art and Future)," *at - Automatisierungstechnik*, vol. 56, no. 1, pp. 4–19, 2008.

[70] G. P. Liu, "Predictive Controller Design of Networked Systems with Communication Delays and Data Loss," *IEEE Transactions on Circuits and Systems II: Express Briefs*, vol. 57, no. 6, pp. 481–485, 2010.

[71] G. P. Liu, J. Mu, and D. Rees, "Networked Predictive Control of Systems With Random Communication Delay," in *UKACC International Conference on Control*, 2004.

[72] G. P. Liu, J. X. Mu, D. Rees, and S. C. Chai, "Design and Stability Analysis of Networked Control Systems with Random Communication Time Delay using the Modified MPC," *International Journal of Control*, vol. 79, no. 4, pp. 288–297, 2006.

[73] G. P. Liu, Y. Xia, J. Chen, D. Rees, and W. Hu, "Networked Predictive Control of Systems with Random Network Delays in Both Forward and Feedback Channels," *IEEE Transactions on Industrial Electronics*, vol. 54, no. 3, pp. 1282–1297, 2007.

[74] Q. Liu, Z. Wang, X. He, and D. Zhou, "A Survey of Event-Based Strategies on Control and Estimation," *Systems Science & Control Engineering*, vol. 2, no. 1, pp. 90–97, 2014.

[75] R. Luck and A. Ray, "An Observer-Based Compensator for Distributed Delays," *Automatica*, vol. 26, no. 5, pp. 903–908, 1990.

[76] J. Lunze, *Regelungstechnik 1*, Berlin Heidelberg: Springer, 2007.

[77] J. Lunze, editor, *Control Theory of Digitally Networked Dynamic Systems*, Springer International Publishing, 2014.

[78] J. Lunze and D. Lehmann, "A State-Feedback Approach to Event-Based Control," *Automatica*, vol. 46, no. 1, pp. 211 – 215, 2010.

[79] W. J. Ma and V. Gupta, "Input-to-state Stability of Hybrid Systems with Receding Horizon Control in the Presence of Packet Dropouts," *Automatica*, vol. 48, no. 8, pp. 1920–1923, 2012.

[80] C. Meng, T. Wang, W. Chou, S. Luan, Y. Zhang, and Z. Tian, "Remote Surgery Case: Robot-Assisted Teleneurosurgery," in *Proc. of the IEEE International Conference on Robotics and Automation (ICRA)*, vol. 1, pp. 819–823, 2004.

[81] M. Moayedi, Y. K. Foo, and Y. C. Soh, "Adaptive Kalman Filtering in Networked Systems With Random Sensor Delays, Multiple Packet Dropouts and Missing Measurements," *IEEE Transactions on Signal Processing*, vol. 58, no. 3, pp. 1577–1588, 2010.

[82] M. Moayedi, Y. K. Foo, and Y. C. Soh, "Filtering for Networked Control Systems with Single/Multiple Measurement Packets subject to Multiple-step Measurement Delays and Multiple Packet Dropouts," *International Journal of Systems Science*, vol. 42, no. 3, pp. 335–348, 2011.

[83] M. Moayedi, Y. K. Foo, and Y. C. Soh, "LQG Control for Networked Control Systems with Random Packet Delays and Dropouts via Multiple Predictive-Input Control Packets," in *Preprints of the IFAC World Congress*, vol. 18, pp. 72–77, 2011.

[84] L. A. Montestruque and P. Antsaklis, "Stability of Model-Based Networked Control Systems With Time-Varying Transmission Times," *IEEE Transactions on Automatic Control*, vol. 49, no. 9, pp. 1562–1572, 2004.

[85] L. A. Montestruque and P. J. Antsaklis, "On the Model-based Control of Networked Systems," *Automatica*, vol. 39, no. 10, pp. 1837–1843, 2003.

[86] J. R. Moyne and D. Tilbury, "The Emergence of Industrial Control Networks for Manufacturing Control, Diagnostics, and Safety Data," *Proc. of the IEEE*, vol. 95, no. 1, pp. 29–47, 2007.

[87] R. Murray, K. Åström, S. Boyd, R. Brockett, and G. Stein, "Future Directions in Control in an Information-Rich World," *IEEE Control Systems Magazine*, vol. 23, no. 2, pp. 20–33, 2003.

[88] P. Naghshtabrizi and J. P. Hespanha, "Designing an Observer-Based Controller for a Network Control System," in *Proc. of the IEEE Conference on Decision and Control and the European Control Conference*, pp. 848–853, 2005.

[89] G. Nair, F. Fagnani, S. Zampieri, and R. Evans, "Feedback Control Under Data Rate Constraints: An Overview," *Proc. of the IEEE*, vol. 95, no. 1, pp. 108–137, 2007.

[90] J. Nilsson, "Real-time Control Systems with Delays," Ph.D. thesis, Ph.D. Dissertation, Department of Automatic Control, Lund Institute of Technology, 1998.

[91] J. Nilsson and B. Bernhardsson, "LQG Control Over a Markov Communication Network," in *Proc. of the IEEE Conference on Decision and Control*, pp. 4586–4591, 1997.

[92] J. Nilsson, B. Bernhardsson, and B. Wittenmark, "Some Topics in Real-time Control," in *Proc. of the American Control Conference*, vol. 4, pp. 2386–2390, 1998.

[93] J. Nilsson, B. Bernhardsson, and B. Wittenmark, "Stochastic Analysis and Control of Real-time Systems with Random Time Delays," *Automatica*, vol. 34, no. 1, pp. 57–64, 1998.

[94] A. Onat, T. Naskali, E. Parlakay, and O. Mutluer, "Control Over Imperfect Networks: Model-Based Predictive Networked Control Systems," *IEEE Transactions on Industrial Electronics*, vol. 58, no. 3, pp. 905–913, 2011.

[95] J. Østergaard and D. E. Quevedo, "Multiple Descriptions for Packetized Predictive Control over Erasure Channels," in *Proc. of the IEEE International Conference on Control and Automation*, pp. 165–170, 2011.

[96] S. Pan, H. Su, H. Wang, and J. Chu, "The Study of Joint Input and State Estimation with Kalman Filtering," *Transactions of the Institute of Measurement and Control*, vol. 33, no. 8, pp. 901–918, 2011.

[97] Z. H. Pang, G. P. Liu, D. Zhou, and M. Chen, "Output Tracking Control for Networked Systems: A Model-Based Prediction Approach," *IEEE Transactions on Industrial Electronics*, vol. 61, no. 9, pp. 4867–4877, 2014.

[98] D. Paret, *FlexRay and its Applications*, New York : Wiley & Sons, 2012.

[99] D. Muñoz de la Peña and P. Christofides, "Lyapunov-Based Model Predictive Control of Nonlinear Systems Subject to Data Losses," *IEEE Transactions on Automatic Control*, vol. 53, no. 9, pp. 2076–2089, 2008.

[100] L. L. Peterson and B. S. Davie, *Computer Networks: A Systems Approach*, New York: Elsevier, 4 edn., 2007.

[101] Phoenix Contact Electronics GmbH, *http://www.interbus.de*

[102] G. Pin and T. Parisini, "Stabilization of Networked Control Systems by Nonlinear Model Predictive Control: A Set Invariance Approach," in *Nonlinear Model Predictive Control*, vol. 384 of *Lecture Notes in Control and Information Sciences*, pp. 195–204, Berlin: Springer, 2009.

[103] G. Pin and T. Parisini, "Networked Predictive Control of Uncertain Constrained Nonlinear Systems: Recursive Feasibility and Input-to-State Stability Analysis," *IEEE Transactions on Automatic Control*, vol. 56, no. 1, pp. 72–87, 2011.

[104] N. J. Ploplys, P. A. Kawka, and A. G. Alleyne, "Closed-loop Control over Wireless Networks," *IEEE Control Systems Magazine*, vol. 24, no. 3, pp. 58–71, 2004.

[105] I. G. Polushin, P. X. Liu, and C. H. Lung, "On the Model-Based Approach to Nonlinear Networked Control Systems," *Automatica*, vol. 44, no. 9, pp. 2409 – 2414, 2008.

[106] PROFIBUS Nutzerorganisation e.V., *http://www.profibus.com*

[107] D. Quevedo and I. Jurado, "Stability of Sequence-Based Control With Random Delays and Dropouts," *IEEE Transactions on Automatic Control*, vol. 59, no. 5, pp. 1296–1302, 2014.

[108] D. Quevedo, J. Østergaard, and D. Nešić, "Packetized Predictive Control of Stochastic Systems Over Bit-Rate Limited Channels With Packet Loss," *IEEE Transactions on Automatic Control*, vol. 56, no. 12, pp. 2854–2868, 2011.

[109] D. Quevedo, E. Silva, and G. Goodwin, "Control over Unreliable Networks Affected by Packet Erasures and Variable Transmission Delays," *IEEE Journal on Selected Areas in Communications*, vol. 26, no. 4, pp. 672–685, 2008.

[110] D. E. Quevedo and D. Nešić, "On Stochastic Stability of Packetized Predictive Control of Non-linear Systems over Erasure Channels," in *IFAC Symposium on Nonlinear Control Systems*, pp. 557–562, 2010.

[111] D. E. Quevedo and D. Nešić, "Input-to-State Stability of Packetized Predictive Control over Unreliable Networks Affected by Packet-Dropouts," *IEEE Transactions on Automatic Control*, vol. 56, no. 2, pp. 370–375, 2011.

[112] D. E. Quevedo and D. Nešić, "Robust Stability of Packetized Predictive Control of Nonlinear Systems with Disturbances and Markovian Packet Losses," *Automatica*, vol. 48, no. 8, pp. 1803–1811, 2012.

[113] D. E. Quevedo, E. I. Silva, and G. C. Goodwin, "Packetized Predictive Control over Erasure Channels," in *Proc. of the American Control Conference*, pp. 1003–1008, 2007.

[114] V. Rao and D. Bernstein, "Naive control of the double integrator," *IEEE Control Systems Magazine*, vol. 21, no. 5, pp. 86–97, 2001.

[115] M. Reble, D. E. Quevedo, and F. Allgöwer, "Control over Erasure Channels: Stochastic Stability and Performance of Packetized Unconstrained Model Predictive Control," *International Journal of Robust and Nonlinear Control*, vol. 23, no. 10, pp. 1151–1167, 2013.

[116] J. P. Richard, "Time-Delay Systems: An Overview of Some Recent Advances and Open Problems," *Automatica*, vol. 39, no. 10, pp. 1667 – 1694, 2003.

[117] M. G. Rivera and A. Barreiro, "Analysis of Networked Contorl Systems with Drops and Variable Delays," *Automatica*, vol. 43, no. 12, pp. 2054 – 2059, 2007.

[118] L. Schenato, "Optimal Estimation in Networked Control Systems Subject to Random Delay and Packet Drop," *IEEE Transactions on Automatic Control*, vol. 53, no. 5, pp. 1311–1317, 2008.

[119] L. Schenato, "To Zero or to Hold Control Inputs With Lossy Links?" *IEEE Transactions on Automatic Control*, vol. 54, no. 5, pp. 1093 –1099, 2009.

[120] L. Schenato, B. Sinopoli, M. Franceschetti, K. Poolla, and S. S. Sastry, "Foundations of Control and Estimation over Lossy Networks," *Proc. of the IEEE*, vol. 95, no. 1, pp. 163–187, 2007.

[121] P. Seiler and R. Sengupta, "Analysis of Communication Losses in Vehicle Control Problems," in *Proc. of the American Control Conference*, vol. 2, pp. 1491–1496, 2001.

[122] P. Seiler and R. Sengupta, "An H_∞ Approach to Networked Control," *IEEE Transactions on Automatic Control*, vol. 50, no. 3, pp. 356–364, 2005.

[123] S. K. Sen, *Fieldbus and Networking in Process Automation*, Boca Raton, Florida: CRC Press, 2014.

[124] Y. Shi and B. Yu, "Output Feedback Stabilization of Networked Control Systems with Random Delays Modeled by Markov Chains," *IEEE Transactions on Automatic Control*, vol. 54, no. 7, pp. 1668–1674, 2009.

[125] H. Shousong and Z. Qixin, "Stochastic Optimal Control and Analysis of Stability of Networked Control Systems with Long Delay," *Automatica*, vol. 39, no. 11, pp. 1877–1884, 2003.

[126] B. Sinopoli, L. Schenato, M. Franceschetti, K. Poolla, M. I. Jordan, and S. S. Sastry, "Kalman Filtering with Intermittent Observations," *IEEE Transactions on Automatic Control*, vol. 49, no. 9, pp. 1453–1464, 2004.

[127] J. J. E. Slotine and W. Li, *Applied Nonlinear Control*, Englewood Cliffs, New Jersey: Prentice Hall, 1991.

[128] O. J. M. Smith, "Closer Control of Loops With Dead Time," *Chemical Engineering Progress*, vol. 53, no. 5, pp. 217–219, 1957.

[129] S. C. Smith and P. Seiler, "Estimation with Lossy Measurements: Jump Estimators for Jump Systems," *IEEE Transactions on Automatic Control*, vol. 48, no. 12, pp. 2163–2171, 2003.

[130] W. Stallings, *Data and Computer Communications*, Englewood Cliffs, New Jersey: Prentice Hall, 9 edn., 2011.

[131] J. F. Sturm, *Using SeDuMi 1.02, A Matlab Toolbox for Optimization over Symmetric Cones*Tech. rep., Tilburg University, Department of Econometrics, 2001.

[132] X. M. Sun, D. Wu, G. P. Liu, and W. Wang, "Input-to-State Stability for Networked Predictive Control With Random Delays in Both Feedback and Forward Channels," *IEEE Transactions on Industrial Electronics*, vol. 61, no. 7, pp. 3519–3526, 2014.

[133] P. Tabuada, "Event-Triggered Real-Time Scheduling of Stabilizing Control Tasks," *Automatica*, vol. 52, pp. 1680–1685, 2007.

[134] P. L. Tang and C. W. de Silva, "Compensation for Transmission Delays in an Ethernet-Based Control Network using Variable-Horizon Predictive Control," *IEEE Transactions on Control Systems Technology*, vol. 14, no. 4, pp. 707–718, 2006.

[135] P. L. Tang and C. W. de Silva, "Stability Validation of a Constrained Model Predictive Networked Control System with Future Input Buffering," *International Journal of Control*, vol. 80, no. 12, pp. 1954–1970, 2007.

[136] G. S. Tian, F. Xia, and Y. C. Tian, "Predictive Compensation for Variable Network Delays and Packet Losses in Networked Control Systems," *Computers & Chemical Engineering*, vol. 39, pp. 152–162, 2012.

[137] Y. C. Tian and D. Levy, "Compensation for Control Packet Dropout in Networked Control Systems," *Information Sciences*, vol. 178, pp. 1263 – 1278, 2008.

[138] Y. Tipsuwan and M. Y. Chow, "Control Methodologies in Networked Control Systems," *Control Engineering Practice*, vol. 11, no. 10, pp. 1099 – 1111, 2003, <ce:title>Special Section on Control Methods for Telecommunication</ce:title>.

[139] S. Trimpe and R. D'Andrea, "Event-Based State Estimation with Variance-Based Triggering," in *Proc. of the IEEE Conference on Decision and Control*, pp. 6583–6590, 2012.

[140] A. Ulusoy, O. Gurbuz, and A. Onat, "Wireless Model-Based Predictive Networked Control System Over Cooperative Wireless Network," *IEEE Transactions on Industrial Informatics*, vol. 7, no. 1, pp. 41–51, 2011.

[141] O. L. do Valle Costa, M. D. Fragoso, and R. P. Marques, *Discrete-Time Markov Jump Linear Systems*, London: Springer, 2005.

[142] D. H. Varsakelis and W. S. Levine, editors, *Handbook of Networked and Embedded Control Systems*, Birkhäuser Boston, 2005.

[143] P. Varutti, B. Kern, T. Faulwasser, and R. Findeisen, "Event-based Model Predictive Control for Networked Control Systems," in *Proc. of the IEEE Conference on Decision and Control*, pp. 567–572, 2009.

[144] G. C. Walsh, H. Ye, and L. G. Bushnell, "Stability Analysis of Networked Control Systems," *IEEE Transactions on Control Systems Technology*, vol. 10, no. 3, pp. 438–446, 2002.

[145] Y. L. Wang and G. H. Yang, "Output Tracking Control for Networked Control Systems with Time Delay and Packet Dropout," *International Journal of Control*, vol. 81, no. 11, pp. 1709–1719, 2008.

[146] Z. Wang, F. Yang, D. W. C. Ho, and X. Liu, "Robust H_∞ Control for Networked Systems With Random Packet Losses," *IEEE Transactions on Systems, Man and Cybernetics - Part B: Cybernetics*, vol. 37, no. 4, pp. 916–924, 2007.

[147] A. Willig, M. Kubisch, C. Hoene, and A. Wolisz, "Measurements of a Wireless Link in an Industrial Environment using an IEEE 802.11-Compliant Physical Layer," *IEEE Transactions on Industrial Electronics*, vol. 49, no. 6, pp. 1265–1282, 2002.

[148] N. van de Wouw, P. Naghshtabrizi, M. B. G. Cloosterman, and J. P. Hespanha, "Tracking Control for Sampled-Data Systems with Uncertain Time-Varying Sampling Intervals and Delays," *International Journal of Robust and Nonlinear Control*, vol. 20, no. 4, pp. 387–411, 2010.

[149] D. Wu, J. Wu, and S. Chen, "Separation Principle for Networked Control Systems with Multiple-Packet Transmission," *IET Control Theory & Applications*, vol. 5, no. 3, pp. 507–513, 2011.

[150] H. Wu, L. Lou, C. C. Chen, S. Hirche, and K. Kuhnlenz, "Cloud-Based Networked Visual Servo Control," *IEEE Transactions on Industrial Electronics*, vol. 60, no. 2, pp. 554–566, 2013.

[151] L. Xiao, A. Hassibi, and J. P. How, "Control with Random Communication Delays via a Discrete-Time Jump System Approach," in *Proc. of the American Control Conference*, vol. 3, pp. 2199–2204, 2000.

[152] R. Yang, G. P. Liu, P. Shi, C. Thomas, and M. V. Basin, "Predictive Output Feedback Control for Networked Control Systems," *IEEE Transactions on Industrial Electronics*, vol. 61, no. 1, pp. 512–520, 2014.

[153] T. C. Yang, "Networked Control System: A Brief Survey," *Control Theory and Applications, IEE Proceedings*, vol. 153, no. 4, pp. 403–412, 2006.

[154] Y. Yang, Y. Wang, and S. H. Yang, "Design of a Networked Control System with Random Transmission Delay and Uncertain Process Parameters," *International Journal of Systems Science*, vol. 39, no. 11, pp. 1065–1074, 2008.

[155] B. Yu, Y. Shi, and Y. Lin, "Discrete-time H_2 Output Tracking Control of Wireless Networked Control Systems with Markov Communication Models," *Wireless Communications and Mobile Computing*, vol. 11, no. 8, pp. 1107–1116, 2011.

[156] M. Yu, L. Wang, G. Xie, and T. Chu, "Stabilization of Networked Contol Systems with Data Packet Dropout via Switched System Approach," in *IEEE International Symposium on Computer Aided Control Systems Design*, pp. 362–367, 2004.

[157] D. Yue, Q. L. Han, and J. Lam, "Network-Based Robust Control of Systems with Uncertainty," *Automatica*, vol. 41, no. 6, pp. 999 – 1007, 2005.

[158] D. Yue, Q. L. Han, and C. Peng, "State Feedback Controller Design of Networked Control Systems," *IEEE Transactions on Circuits and Systems II: Express Briefs*, vol. 51, no. 11, pp. 640–644, 2004.

[159] S. Yüksel and T. Başar, *Stochastic Networked Control Systems*, New York: Springer, 2013.

[160] S. Yüksel, H. Hindi, and L. Crawford, "Optimal Tracking with Feedback-Feedforward Control Separation over a Network," in *Proc. of the American Control Conference*, pp. 3500–3506, 2006.

[161] L. Zhang, H. Gao, and O. Kaynak, "Network-Induced Constraints in Networked Control Systems - A Survey," *IEEE Transactions on Industrial Informatics*, vol. 9, no. 1, pp. 403–416, 2013.

[162] L. Zhang, Y. Shi, T. Chen, and B. Huang, "A New Method for Stabilization of Networked Control Systems with Random Delays," *IEEE Transactions on Automatic Control*, vol. 50, no. 8, pp. 1177–1181, 2005.

[163] W. Zhang, M. S. Branicky, and S. M. Phillips, "Stability of Networked Control Systems," *IEEE Control Systems Magazine*, vol. 21, no. 1, pp. 84–99, 2001.

[164] Y. B. Zhao, Y. Kang, G. P. Liu, and D. Rees, "Stochastic Stabilization of Packet-Based Networked Control Systems," *International Journal of Innovative Computing, Information and Control*, vol. 7, no. 5, pp. 2441–2455, 2011.

[165] Y. B. Zhao, J. Kim, and G. P. Liu, "Error Bounded Sensing for Packet-Based Networked Control Systems," *IEEE Transactions on Industrial Electronics*, vol. 58, no. 5, pp. 1980–1989, 2011.

[166] Y. B. Zhao, J. Kim, and G. P. Liu, "Offline Model Predictive Control-Based Gain Scheduling for Networked Control Systems," *IET Control Theory & Applications*, vol. 6, no. 16, pp. 2585–2591, 2012.

[167] Y. B. Zhao, G. P. Liu, and D. Rees, "Design of a Packet-Based Control Framework for Networked Control Systems," *IEEE Transactions on Control Systems Technology*, vol. 17, no. 4, pp. 859–865, 2009.

[168] Y. B. Zhao, G. P. Liu, and D. Rees, "Packet-Based Deadband Control for Internet-Based Networked Control Systems," *IEEE Transactions on Control Systems Technology*, vol. 18, no. 5, pp. 1057–1067, 2010.

[169] Y. Zheng, H. Fang, and H. O. Wang, "Takagi-Sugeno Fuzzy-Model-Based Fault Detection for Networked Control Systems with Markov Delays," *IEEE Transactions on Systems, Man, and Cybernetics, Part B: Cybernetics*, vol. 36, no. 4, pp. 924–929, 2006.

Own Publications

[170] M. Dolgov, J. Fischer, and U. D. Hanebeck, "Event-based LQG Control over Networks Subject to Random Transmission Delays and Packet Losses," in *Proc. of the IFAC Workshop on Distributed Estimation and Control in Networked Systems*, 2013.

[171] M. Dolgov, J. Fischer, and U. D. Hanebeck, "Sequence-based LQG Control with Linear Integral Constraints over Stochastic Networks (to appear)," in *Proc. of the IEEE Conference on Decision and Control*, 2014.

[172] J. Fischer, M. Dolgov, and U. D. Hanebeck, "On Stability of Sequence-Based LQG Control," in *Proc. of the IEEE Conference on Decision and Control*, 2013.

[173] J. Fischer, M. Dolgov, and U. D. Hanebeck, "Optimal Sequence-Based Tracking Control over Unreliable Networks," in *Proc. of the IFAC World Congress*, 2014.

[174] J. Fischer, A. Hekler, M. Dolgov, and U. D. Hanebeck, "Optimal Sequence-Based LQG Control over TCP-like Networks Subject to Random Transmission Delays and Packet Losses," in *Proc. of the American Control Conference*, pp. 1543–1549, 2013.

[175] J. Fischer, A. Hekler, and U. D. Hanebeck, "State Estimation in Networked Control Systems," in *Proc. of the International Conference on Information Fusion*, pp. 1947–1954, 2012.

[176] J. Fischer, M. Reinhardt, and U. D. Hanebeck, "Optimal Sequence-Based Control and Estimation of Networked Linear Systems," *at - Automatisierungstechnik*, vol. 61, no. 7, pp. 467–476, 2013.

[177] L. Grüne, F. Allgöwer, R. Findeisen, J. Fischer, D. Groß, U. D. Hanebeck, B. Kern, M. Müller, J. Pannek, M. Reble, O. Stursberg, P. Varutti, and K. Worthmann, "Distributed and Networked Model Predictive Control," in J. Lunze, editor, *Control Theory of Digitally Networked Dynamic Systems*, pp. 111–167, Springer International Publishing, 2014.

[178] A. Hekler, J. Fischer, and U. D. Hanebeck, "Control over Unreliable Networks Based on Control Input Densities," in *Proc. of the International Conference on Information Fusion*, 2012.

[179] A. Hekler, J. Fischer, and U. D. Hanebeck, "Sequence-Based Control for Networked Control Systems Based on Virtual Control Inputs," in *Proc. of the IEEE Conference on Decision and Control*, pp. 7–13, 2012.

Karlsruhe Series on Intelligent Sensor-Actuator-Systems

Edited by Prof. Dr.-Ing. Uwe D. Hanebeck // ISSN 1867-3813

Band 1 **Oliver Schrempf**
Stochastische Behandlung von Unsicherheiten in kaskadierten dynamischen Systemen. 2008
ISBN 978-3-86644-287-0

Band 2 **Florian Weißel**
Stochastische modell-prädiktive Regelung nichtlinearer Systeme. 2009
ISBN 978-3-86644-348-8

Band 3 **Patrick Rößler**
Telepräsente Bewegung und haptische Interaktion in ausgedehnten entfernten Umgebungen. 2009
ISBN 978-3-86644-346-4

Band 4 **Kathrin Roberts**
Modellbasierte Herzbewegungsschätzung für robotergestützte Interventionen. 2009
ISBN 978-3-86644-353-2

Band 5 **Felix Sawo**
Nonlinear state and parameter estimation of spatially distributed systems. 2009
ISBN 978-3-86644-370-9

Band 6 **Gregor F. Schwarzenberg**
Untersuchung der Abbildungseigenschaften eines 3D-Ultraschall-Computertomographen zur Berechnung der 3D-Abbildungsfunktion und Herleitung einer optimierten Sensorgeometrie. 2009
ISBN 978-3-86644-393-8

Die Bände sind unter www.ksp.kit.edu als PDF frei verfügbar oder als Druckausgabe bestellbar.

Karlsruhe Series on Intelligent Sensor-Actuator-Systems

Edited by Prof. Dr.-Ing. Uwe D. Hanebeck // ISSN 1867-3813

Band 7 **Marco Huber**
Probabilistic Framework for Sensor Management. 2009
ISBN 978-3-86644-405-8

Band 8 **Frederik Beutler**
Probabilistische modellbasierte Signalverarbeitung zur instantanen Lageschätzung. 2010
ISBN 978-3-86644-442-3

Band 9 **Marc Peter Deisenroth**
Efficient Reinforcement Learning using Gaussian Processes. 2010
ISBN 978-3-86644-569-7

Band 10 **Evgeniya Ballmann**
Physics-Based Probabilistic Motion Compensation of Elastically Deformable Objects. 2012
ISBN 978-3-86644-862-9

Band 11 **Peter Krauthausen**
Learning Dynamic Systems for Intention Recognition in Human-Robot-Cooperation. 2013
ISBN 978-3-86644-952-7

Band 12 **Antonia Pérez Arias**
Haptic Guidance for Extended Range Telepresence. 2013
ISBN 978-3-7315-0035-3

Band 13 **Marcus Baum**
Simultaneous Tracking and Shape Estimation of Extended Objects. 2013
ISBN 978-3-7315-0078-0

Karlsruhe Series on Intelligent Sensor-Actuator-Systems

Edited by Prof. Dr.-Ing. Uwe D. Hanebeck // ISSN 1867-3813

Band 14 **Benjamin Noack**
State Estimation for Distributed Systems with Stochastic and Set-membership Uncertainties. 2014
ISBN 978-3-7315-0124-4

Band 15 **Jörg Fischer**
Optimal Sequence-Based Control of Networked Linear Systems. 2014
ISBN 978-3-7315-0305-7